D1047277

Developmental Teaching of MATHEMATICS for the LEARNING DISABLED

Contributors

Colleen S. Blankenship

Anne Marie Fitzmaurice-Hayes

Henry A. Goodstein

Elizabeth McEntire

Mahesh C. Sharma

Robert A. Shaw

Raymond E. Webster

Developmental Teaching of MATHEMATICS for the LEARNING DISABLED

Editor

John F. Cawley

University of New Orleans
Lakefront

AN ASPEN PUBLICATION®
Aspen Publishers, Inc.
Rockville, Maryland
Royal Tunbridge Wells
1984

Library of Congress Cataloging in Publication Data
Main entry under title:

Developmental teaching of mathematics for the learning disabled.

"An Aspen publication."
Includes index.
1. Mathematics — Study and teaching. 2. Learning disabilities.
I. Cawley, John F.
QA11.D44 1983 371.9'044 84-6247
ISBN: 0-89443-581-7

Publisher: John Maroszan
Associate Publisher: Jack W. Knowles, Jr.
Editorial Director: Margaret M. Quinlin
Editor-in-Chief: Michael Brown
Executive Managing Editor: Margot G. Raphael
Managing Editor: M. Eileen Higgins
Editorial Services: Scott Ballotin
Printing and Manufacturing: Debbie Collins

Library of Congress Catalog Card Number: 84-6247
ISBN: 0-89443-581-7

Printed in the United States of America

3 4 5

Table of Contents

Preface

Mathematics is a broad topic to which children with learning disabilities are exposed from infancy through their developmental years. The acquisition of skills and concepts and the use of mathematics is not limited to the school years, nor, for that matter, to any period of life. Although certain aspects of mathematics for the learning disabled receive more attention in school, a wide variety of concepts and skills will be needed throughout life.

Children with learning disabilities show severe discrepancies or disorders in one or more of a combination of academic and/or cognitive traits. With respect to mathematics, these traits may be in skills or in problem solving and reasoning or in the cognitive abilities that may limit performance in mathematics. Children with learning disabilities may also manifest deficiencies or deficits in areas such as language or reading comprehension or in other qualities and these may inhibit the development of mathematics concepts or skills.

Written by specialists in each area, *Developmental Teaching of Mathematics for the Learning Disabled* is a presentation of selected facets of mathematics and mathematics applications from preschool through the adult years. This progression recognizes that learning-disabled individuals attend school for some 12 to 14 or more years. Thus, there is a commitment to both curriculum and instruction and to the development and implementation of both as necessary components of an appropriate education.

Chapter 1 describes selected qualities of the concept of learning disability. The overriding premise is that this concept represents a heterogeneous phenomenon that has different meanings and implications for different people.

Chapter 2 brings together the topics of curriculum and instruction. Curriculum, the "what shall we teach, when shall we teach it, and in what sequence is it best taught," has generally received less attention from special educators than instruction, "how shall we teach it." In this chapter both curriculum and instruction are seen as partners in the education of the learning disabled.

Effective assessment is fundamental to appropriate educational placement based on the identification of present levels of functioning and the development of goals and objectives. There are numerous approaches to assessment at various developmental levels. Many of these are discussed in Chapter 3.

Chapter 4 develops the premise that the needs of learning-disabled children are interrelated and that activities from one topic or skill area can be used to reinforce desirable behaviors or to promote opportunities for application and generalization.

Chapters 5 through 8 contain extensive presentations of curriculum topics and instructional options. The continuity expressed across these chapters is representative of the continuity desirable in programs for the learning disabled.

Chapter 9 branches to an exploration of mathematics in day-to-day living. The need for mathematical understanding and skill is pervasive and our efforts must be directed toward preparation for knowledgeable participation.

Chapter 10 discusses the identification, selection, and modification of instructional materials. Two major points are stressed. The first is that one should have some model or framework upon which to base selections or to elect modifications. The second is that modifications should have an internal consistency and offer diverse representations of a given topic or skill.

The topic of classroom management is likely to be of greater concern than almost any other issue. Questions such as "How do I help the learning disabled when I have 25 other children?" or "How do I develop acceptable behaviors in these children?" are often asked. Chapter 11 contains a comprehensive review of approaches to classroom management at varying developmental levels.

Chapter 12 is designed to provide an understanding of the administrative options and models that are used to deliver services to the learning disabled. The chapter differentiates the elementary and secondary levels and provides suggestions as to how to integrate mathematics with other topics.

Developmental Teaching of Mathematics for the Learning Disabled is comprehensive in its overview and format. Its aim is to aid in the selection among curriculum, instructional, and management options that are available for the local education agency.

Clearly, there is a need for cooperation among all members of the community. Mathematics educators and special educators must work together. Curriculum and instructional specialists must pool their expertise. Joint goals and joint efforts will produce programs that will develop the skills and applications needed by the learning-disabled population we serve.

John F. Cawley
Editor

Chapter 1

Learning Disabilities: Issues and Alternatives

John F. Cawley

The term *learning disability* is used throughout this book to express a concept. As such, it is an organizer or heading under which traits or qualities may be grouped according to attributes or rules. To view learning disability as a concept is to recognize its diversity, a diversity that is especially noticeable when one attempts to assemble the characteristics of individuals who are referred to as learning disabled. This is so because individuals vary in their characteristics and because the teachers, clinicians, and research workers who serve and study the learning disabled often have different perspectives (Harber, 1981; Kavale & Nye, 1981; Olson & Mealor, 1981).

THE FIELD OF LEARNING DISABILITIES: ISSUES

Several trends in the field of learning disability appear to be related to this diversity. The first is that no matter what the topic or the methodology, studies generally are showing performance levels of the learning disabled below those of the non-learning disabled. This is true in mathematics, reading, spelling, and in process comparisons such as conceptual learning.

A second trend shows that the characteristics or traits that are being ascribed to the learning disabled are a result of the diagnostic approach or research activity employed with these individuals. Whether we have assessed or researched the correct or most appropriate set of characteristics remains to be seen.

One study (Olson & Mealor, 1981) showed that 55 percent of 113 studies reviewed used academics for subject identification yet only half of these described academic problems in the context of a significant discrepancy. Only 23 percent of these studies included reference to cognitive processes, ample evidence of the selectivity in research.

A third trend is related to remediation and treatment. Within this limited area, even with the most experimental of treatments, some individuals are making more progress than others and some make little if any progress. To date, we do not seem able to predict how much progress an individual will make in a given amount of treatment or, put another way, we do not seem to know how long it will take a given individual to attain an expected level of proficiency.

Swanson (1982) studied conceptual process as a function of age and enforced attention. Comparisons between learning-disabled and non-learning-disabled subjects demonstrated a more favorable level of performance for the non-learning disabled. However, several points should be kept in mind: (1) some learning-disabled youngsters solved the most difficult of problems, (2) only one 45-minute training session was provided, and (3) the error rate decreased in both groups as age increased. It would be of value to know what characteristics distinguished the more able from the less able among the learning-disabled sample, whether more than one training session would accentuate performance, and if there is an age at which the learning disabled master specific conceptual processes.

Lloyd, Saltzman, and Kauffman (1981) examined generalization in academic learning as a result of preskill and strategy training in multiplication and division. The training included rote counting by a given number (e.g., 5), work with a trainer to develop proficiency with preskills, strategy training involving a counting methodology, and cue training in which the individual was shown counting sequences for numbers not taught during preskills. This was followed by a second experiment in which additional strategy training was provided and generalizations were observed. The authors concluded that it is possible to teach learning-disabled children academic skills if they are given adequate instruction. They may not have to be taught every item in each skill area and can transfer training appropriately. Generalization in academic learning may be predictable on the basis of the instruction provided. The results of this study clearly have implications for the tendency to teach items strictly by rote to learning-disabled children.

By any measure, the field of learning disability is anything but static. Since its somewhat formal inception in the 1960s, there have been definitions and changes in definitions; there have been regulations and changes in regulations; there have been theoretical models and changes in these models; and, there have been diagnostic and instructional programs and changes in these programs.

This is a book about mathematics and learning disability. It contains elements of controversy, logical interpretations of the state of the art, and a variety of suggestions relative to curriculum instruction and programming. Some fundamental premises guide the development of this book. They are:

- The child with a learning disability or combination of learning disabilities is entitled to every opportunity to attain his or her maximum potential in mathematics and in the application and enjoyment of mathematics.
- Mathematics must be capable of doing more for an individual than providing only mathematical skills. Experiences with mathematics need to be designed

to impact language, cognition, and the social and other developmental needs of the person.

- Performance discrepancies in mathematics may be due to a number of conditions.
- Some individuals with learning disabilities manifest problems in mathematics and many other areas (e.g., reading, writing); some individuals show problems only in mathematics; some individuals have no problems in mathematics.
- Interpretations of what constitutes a learning disability in general and a learning disability in mathematics vary.
- Assessment techniques may produce conflicting results.

Characteristics of the Learning Disabled

Reisman (1982) suggests that the following conditions may account for performance discrepancies in mathematics:

- gaps in mathematical foundations
- lack of readiness
- emotional problems
- deprived environments
- poor teaching
- learning disabilities

She lists the following as learning disability traits:

- inefficient searching strategies
- inability to retrieve labels
- inability to produce written responses
- poor short-term and/or long-term memory
- poor spatial relations
- impaired communication
- distractibility and excessive fatigability

Glennon and Cruickshank (1981) discuss the characteristics of the learning disabled as related to mathematics within the following groupings:

- attention disturbances and attention span
- figure-ground pathology

- dissociation
- memory dysfunctions
- sequencing
- discrimination
- directionality and body parts
- spatial and temporal disorientation
- obtaining closure
- perseveration
- intersensory disorganization

Johnson (1979) lists the following as learning disability characteristics having relevance to arithmetic:

- memory disabilities
- visual and auditory discrimination disabilities
- visual and auditory association disabilities
- perceptual-motor disabilities
- spatial awareness and orientation disabilities
- verbal expression disabilities
- closure and generalization disabilities
- attending disabilities

What is true with regard to the general relationship between learning disabilities and mathematics is also relevant to the specific characteristics of the learning disabled and their mathematics performance. That is, there are some children with discrepancies, for example, in attending disabilities who have difficulties in mathematics and there are some who do not; there are also some children with difficulties in mathematics who do not have difficulties in attending.

Performance of the Learning Disabled

Pieper and Deshler (1980) contrasted the performance of adolescents having specific arithmetic disabilities with those having specific reading disabilities. Individuals with specific arithmetic disabilities were two or more years lower in arithmetic than reading. Those with specific reading disabilities were two or more years lower in reading than in arithmetic. The range of discrepancy for the arithmetic groups was 2.5 to 8.8 years, far greater than the 1.6 to 4.6 for the reading cases. Of particular relevance to the present discussion was the fact that different tests yielded different discrepancies. If, for example, test A showed the individual to be two years behind in arithmetic, test B showed the same individual three years behind. This made stability in identification difficult. As a result, one should err on the side of the greater discrepancy and place the level of functioning of the individual at the lower level. Curriculum placement would thus be at an easier rather than a more difficult level.

Meltzer (1978) studied abstract reasoning in a group of perceptually handicapped learning-disabled children. Eleven tasks covered conservation of numbers, seriation, classification, and conservation of quantity. Comparisons were made between learning-disabled and non-learning-disabled children and the mode of difficulty of the tasks for the two groups was examined. Learning-disabled children were significantly inferior to non-learning-disabled children in visual-perceptual performance. However, on the 11 performance tasks, no significant differences were found. The order of difficulty of tasks was the same for both groups; those tasks that were difficult for the learning disabled were also difficult for the non-learning disabled.

Greenes (1979) suggests that poor visual perception skills generally lead to poorly developed concepts of space and resulting problems in sequencing, directionality, measurement, geometric forms, and conservation of quantity. This highlights some of the difficulty we experience in attempting to generalize information. Greenes described certain problems of the learning disabled. Meltzer studied some of these and does not substantiate Greenes' observations.

Fleischner, Garnett, and Shepherd (1980) studied the performance of learning-disabled youngsters in arithmetic. Table 1–1 is constructed from their data. Each group was administered a series of three-minute tests of basic facts, 98 addition, 98 subtraction, and where appropriate, 98 multiplication. The numerator represents the number correct and the denominator represents the number attempted. Proficiency, the number attempted divided into the number correct, is high. That is, of the items attempted, the percent correct is high for both the learning disabled and non-learning disabled.

Learning-disabled (LD) children do not seem to be "wild guessers" or random responders. If, in fact, the smaller number attempted was due to the fact that they "passed" on those they did not know, their performance is even more qualitative. If, on the other hand, they were not quick enough to attempt more items in the prescribed time, their performance may not be due to an arithmetical difficulty, per se, but to a problem of speed or on-task behavior.

Performance, both in terms of number attempted and number correct, increased with age. If LD children continue to lag and manifest some degree of discrepancy, attention to procedures that increase their rate of performance are needed or curriculum decisions need to be made to determine what will be dropped in order to provide for the time needed in computation.

Table 1–1 Basic Arithmetic Performance of Learning-Disabled and Non-Learning-Disabled Children

	Addition	*Subtraction*	*Multiplication*
Grade 3			
LD	26/29	18/22	
NLD	57/58		
Grade 4			
LD	35/37	25/28	
NLD	70/71	52/54	
Grade 5			
LD	46/47	33/35	26/32
NLD	77/79	62/64	73/75
Grade 6			
LD	59/60	44/47	42/47
NLD	85/86	73/74	81/82

Source: Fleischner, J., Garnett, K., & Shepherd, M. (1980). *Proficiency in arithmetic basic fact computation of learning disabled and non-disabled children* (Technical Report No. 9). New York: Teachers College, Columbia University, Research Institute for the Study of Learning Disabilities. Used by permission.

Discrepancy or Disorder?

The study by Fleischner et al. raises two additional issues. One is the determination of the difference between a discrepancy and a disorder. A second relates to the prevalence of disorders of thinking.

It seems important to distinguish a discrepancy or difficulty from a disorder, particularly in mathematics. A difficulty tends to suggest trouble in establishing a rate of performance or growth consistent with others of similar ability. A person may experience difficulty within mathematics. This difficulty may slow the person down or may make it impossible for the person to develop certain skills or principles. A disorder, on the other hand, implies an aberration. It is not a case of one not being able to do something; it is a problem in which one does not do something in a manner that is rational or comprehensible.

We can see how a person might not develop skill with a conventional algorithm such as

$$\begin{array}{r} 504 \\ -\ 237 \\ \hline \end{array}$$

at an age when most others do it, but it is extremely difficult to understand how a person is able to habituate a disordered algorithm such as

$$\begin{array}{r} 504 \\ -\ 237 \\ \hline 3 \end{array}$$

where the figures in each row are added (i.e., $5 + 0 + 4$) and then the larger number is subtracted from the smaller number (i.e., $5 + 0 + 4 = 9; 2 + 3 + 7 = 12; 12 - 9 = 3$). For this example, Carpenter, Corbitt, Kepner, Lindquist, and Reys (1980) found the results shown in Table 1–2.

As can be seen, the percent correct increases with age for both the basic facts and the subtraction illustration. This points out the necessity for special educators to understand the development of non-learning-disabled children and to understand curriculum expectancies. Obviously, proficiency with 504 − 237 cannot be expected by age nine, nor can it be totally expected by age thirteen.

Loring (1981) adds an interesting dimension to our topic. He conducted a study of mathematical thinking among learning-disabled and non-learning-disabled children. The samples were: 27 learning-disabled fourth graders, 27 non-learning-disabled fourth graders, and 27 non-learning-disabled third graders. The tasks included mental addition, estimation, understanding of reciprocity and commutativity, counting by tens, several tasks assessing knowledge and skill with

Table 1–2 Percent Correct on Basic Facts

Skill	*Age 9*	*Age 13*	*Age 17*
Addition	89	95	97
Subtraction	79	93	95
Multiplication	60	93	93
Division	—	81	89

Subtraction Computation
"Subtract 237 from 504"

Responses	*Age 9*	*Age 13*
Correct	28	73
Reversal Error	32	5
Regrouping Error	4	7
Other	29	14
"I don't know" or no response	7	1

Source: Carpenter, T., Corbitt, M., Kepner, H., Lindquist, M., & Reys, R. (1980). Results and implications of the second NEAP mathematics assessment: Elementary school. *Arithmetic Teacher, 27,* 10–47. Used by permission.

large numbers, addition facts, place value, written addition and subtraction, monitoring errors on written calculations, and story problems.

The learning disabled performed similar to third graders on all tasks and differed from the fourth graders on only two large number tasks, complex renaming, and complex story problems. Where differences did occur, they were attributed to a subgroup of slightly less than one-half the learning-disabled sample. There was no evidence of unusual thought patterns or qualitative differences. In effect, LD was a discrepancy, not a disorder.

Lepore (1979) contrasted the performance of mildly retarded and learning-disabled children at two different age levels in a number of areas of computation (see Table 1–3). This included an inquiry into the method used by the individual when doing the required tasks. The study is of particular interest because it provides both discrepancy and disorder analysis. In general, the developmental or discrepancy information showed:

1. Young mildly retarded children made more error types in addition than older mildly retarded or younger or older learning-disabled children.
2. Older (CA 160 months) LD children had mastered all the addition items.
3. Subtraction errors were extensive across all groups.
4. Multiplication errors were extensive across all groups.
5. Division errors varied.

The following illustrate the irrational or strange thinking that certain individuals used in explaining their performance.

Table 1–3 Percentage of Multiplication Errors

Error Type	*YMR*	*OMR*	*YLD*	*OLD*
Renaming	11.69	14.91	22.44	6.98
Multiplies by 1 digit only	1.06	5.26	8.16	2.33
Multiplies by wrong number	56.38	28.06	44.9	30.24
Add errors	2.13	10.53	—	6.98
Zeros	9.57	8.78	—	13.95
Facts	14.89	20.18	10.20	30.23
Partial sums	3.19	7.02	—	6.98
Reversals	—	2.63	—	—
Process	1.06	2.63	14.29	—
Total number	94	114	49	43

Source: Lepore, A. (1979). A comparison of computational errors between educable mentally handicapped and learning disabled children. *Focus on Learning Problems in Mathematics, 1*(1), 12–33. Used by permission.

Addition

The problem read:

$$\begin{array}{r} 35 \\ +\ 92 \\ \hline \end{array}$$

To obtain the answer the student said that if we carry we have to cross out. He crossed out the 5, added a 1 next to it then said 5 + 1 = 6 and wrote the 6 above the 5. He then said 1 + 2 = 3 and wrote the 3 in the ones column of the answer. When asked why he added 1 and 2 he replied that the one was "carried." He then wrote 1 next to the 3 and a 2 above it. He then added 9 + 1 = 10, wrote 1 in the tens column and carried the 0.

Subtraction

The subtraction in this problem is done correctly, but two-digit numbers are written backward. The learner claims he writes them backward because he is left handed.

Multiplication

For both problems, this student said that "you can't do it with zero, so I act like it's not there."

Division

The zero is ignored for division purposes, but sometimes is used for multiplication. Four into 9 is 2, put 2 over 9 because 4 divided into the 9. Two times 40 is 80. Subtracts. Four into 17 is 4. 4 × 4 = 16.

Disorders may be attributed to a number of factors.

1. The individual may not be ready for the task when it is taught. The individual does not understand the concept or the algorithm because it is too difficult for his or her stage of development.
2. The individual may have been allowed to habituate an inappropriate concept or algorithm because performance was scored rather than analyzed. For example:

$9\overline{)819}$ (quotient: 1) The individual says: "Nine into nine is one."

$9\overline{)819}$ (quotient: 91) and "Nine into 81 is nine." And the response is correct.

However, when given:

9)839 The individual says: "Nine into nine is one,"

9)839 and "Nine into 83 is nine and two left over."

3. The thinking and reasoning capabilities of the individual are distorted.

A general approach to the matter of disorders is to intervene early and prevent any behavior from being habituated. The use of manipulatives is most helpful in these instances.

A general approach to the matter of discrepancies is to provide a comprehensive program with equivalent representation from many areas of mathematics (e.g., geometry, measurement) and to stress the development of principles and rules as prerequisites or companions with computation, integrating all of these with applications and problem solving.

Definition of Learning Disability

The definition sponsored by the National Joint Committee for Learning Disabilities (Hammill, Leigh, McNutt, & Larsen, 1981) makes no provision for the traditional psychological-perceptual processes such as those discussed previously and contained in the regulations to Public Law 94-142, the Education for All Handicapped Children Act.

Joint Committee

> Learning disabilities is a generic term that refers to a heterogeneous group of disorders manifested by significant difficulties in the acquisition and use of listening, speaking, reading, writing, reasoning or mathematical abilities. These disorders are intrinsic to the individual and presumed to be due to central nervous systems dysfunction. Even though a learning disability may occur concomitantly with other handicapping conditions (e.g., sensory impairment, mental retardation, social and emotional disturbance) or environmental influences (e.g., cultural differences, insufficient/inappropriate instruction, psychogenic factors) it is not the direct result of these conditions or influences. (Hammill et al.)

PL 94-142

> "Specific learning disability" means a disorder in one or more of the basic psychological processes involved in understanding or in using

> language, spoken or written, which may manifest itself in an imperfect ability to listen, think, speak, read, write, spell, or to do mathematical calculations. The term includes such conditions as perceptual handicaps, brain injury, minimal brain dysfunction, dyslexia, and developmental aphasia. The term does not include children who have learning problems which are primarily the result of visual, hearing, or motor handicaps, of mental retardation, or of environmental, cultural, or economic disadvantage. (P.L. 94-142, the Education for All Handicapped Children Act, 1975)

Although these definitions vary somewhat and are discussed more fully elsewhere (Myers & Hammill, 1982), each does contain essentially the same performance areas.

PL 94-142	*NJCLD*
Listen	Listening
Speak	Speaking
Read	Reading
Write	Writing
Mathematical calculations	Mathematical abilities
Think	Reasoning

The listing of individual items as shown above results in a tendency to view individuals in terms of each item. An individual researcher might be attracted to an item such as listening and neglect to examine listening as it relates to each of the other items. For this reason, it seems appropriate to examine the concept of learning disability within the framework of a Venn diagram (see Figure 1–1).

The Venn diagram demonstrates that some aspects of learning disability exist independently whereas others intersect in varying degrees and combinations. This format encourages us to examine the extent to which growth and development in one area may assist or impair growth and development in others. In its broad sense, this has been one of the more delicate issues in the field of learning disability. That is, can growth and development in one area assist or influence growth in others? There have generally been two approaches to this question. The direct approach concentrates instruction in a basic area or on a specific skill (e.g., computation) and attempts to ameliorate problems or to remediate deficiencies through instruction that is specific to the problem. The indirect approach attempts to provide experiences in selected areas (e.g., perceptual training) in the hope that this training will result in improvement in another area (e.g., computation).

Which is the better approach? True, many indirect approaches may not have produced the desired impact. Then again, many direct approaches have not been proven to be panaceas either. Is it possible that we have not considered the correct

Figure 1–1 Venn Diagram of Learning Disabilities

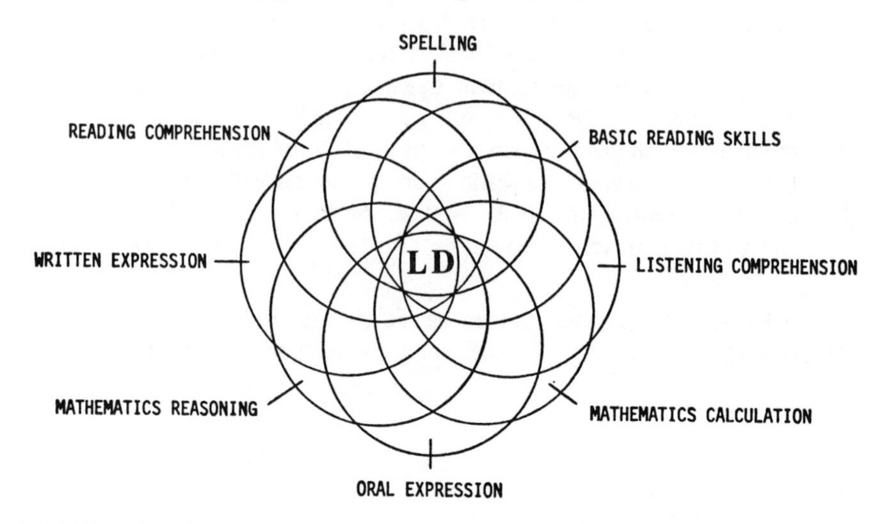

indirect approach (e.g., metacognition training) or a more correct direct approach (e.g., discovery)?

Possibly, another perspective can be beneficial. This perspective suggests the following:

- Effective direct instruction needs to be conducted using appropriate curriculum. The topics chosen must be appropriate for the learner; they must be presented when the learner has attained a satisfactory experiential background and level of readiness; direct instruction needs to be flexible and capable of controlling adaptations in instructional practices and materials (see Chapter 9). Programming needs to be based upon a valid assessment.
- Process or mediational deficits can generally be compensated for by adapting instruction or materials. Short-term memory problems can be adjusted by increasing the number of repetitions during original learning and by practicing items after they have been learned. If, for example, as Webster (1981) points out, there is a loss of 30–40 percent within a few minutes, the teacher could return to the items within a lesser time period. If distractibility and off-task behavior characterize the learner, the teacher could implement procedures to reduce the distractibility and off-task behavior.
- Individuals with processing or mediational deficits, particularly in areas such as thinking and reasoning, can be provided with relevant experiences within the context of mathematics (see Figure 1–2). The purpose in using the content of mathematics is to make these experiences as natural as possible within the

Figure 1–2 Cognitive Activities Using Mathematics Content

Classification

See these. Mark all of these that are the same.

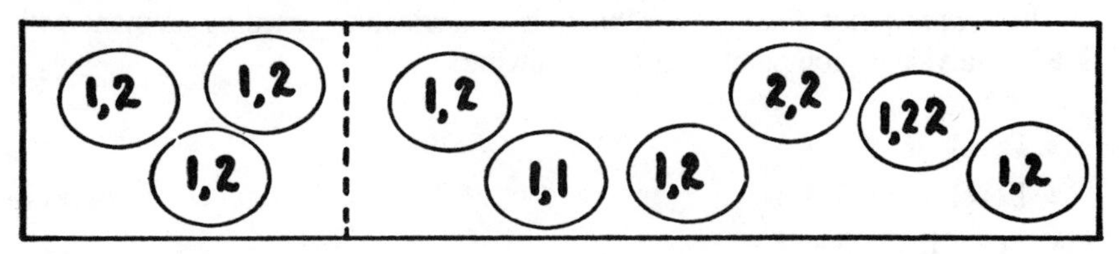

Oddity

See these. Mark one that does not belong.

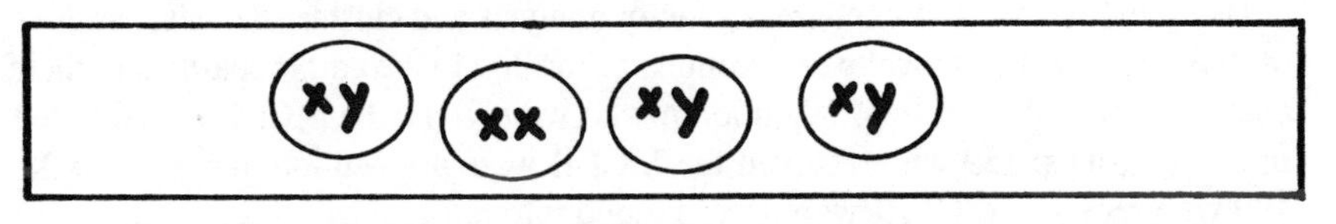

set of ongoing activities. No attempt is made to "remediate" deficiencies. Rather, the intent is to prevent their exclusion because "these kids can't do math" and add to the cumulative deficit. The one thing we do not know about learning-disabled children is what should be excluded from their natural program.

Heterogeneity

As was stated in the beginning of this chapter, considerable heterogeneity exists in the characteristics and patterns of arrangements of these characteristics among learning-disabled individuals.

One study (Cawley, Fitzmaurice, Shaw, Kahn, & Bates, 1979) assembled data on 783 school-identified learning-disabled youth. These students were divided into groups according to level of mathematics assessment. Three levels of the *Mathematics Concept Inventory* (MCI) were administered along with the following:

- *Wechsler Intelligence Scale for Children-R*
- *Wide Range Achievement Test*
- *Peabody Picture Vocabulary Test*
- *Ravens Progressive Matrices Test*

- *Graham-Kendall Memory for Designs Test*
- *Informal Word Recognition Test*
- *Social-Emotional Rating Scale*

The *Mathematics Concept Inventory* is a curriculum-based assessment with grade equivalents comparable to the following:

- Level I: K –1.5
- Level II: 1.5–3.0
- Level III: 3.0–4.5
- Level IV: 4.5–6.0

Only the first three levels were administered.

Using the statistical procedures of factor analysis and cluster analysis, samples of individuals in each level were grouped according to the commonalities in their characteristics. A sample of 39 children who were given the MCI I were divided into five groups; 159 who received the MCI II were divided into ten groups; the 199 who received MCI III were also divided into ten groups.

Note the variation in characteristics as well as the variables that entered into the equations to specify these characteristics. For example, at level I, the Mathematics did not enter into the equation, whereas the subscales of Sets, Patterns, and Measurement were included in level II and the total score entered in at level III.

Bates (1981) intensified the study of heterogeneity among learning-disabled children by conducting an extensive cluster analysis examination of WISC-R subtest performance (see Exhibit 1–1). The eight cluster profiles summarized in Exhibit 1–1 do not show any specific pattern.

> Clusters I, II, III, and V support the notion that large verbal-performance discrepancies are indicative of L.D. children. Clusters I, II, and V represent what Clements and Peters (1962) labeled a pure L.D. profile (performance verbal). Cluster II represents what these researchers labeled a pure hyperkinetic profile (verbal performance). Clusters III, V, VI, and VIII support Bannatyne's suggested pattern (spatial conceptual sequential). Cluster VII lends some support to Swartz's (1974) ACID profile. This support is weakened, however, because cluster VII is characterized by a low block design score, and because digit span scores were not included in this study. Tabachnick's (1979) suggestion that a low coding score is indicative of learning disabilities is clearly supported by clusters VI and VIII. Cluster VIII, however, is also characterized by a poor vocabulary performance. Only cluster IV supports the popular notion that the *WISC-R* profiles of L.D. children

Exhibit 1–1 Summary of the Characteristics of the Eight Cluster Profiles

I. 31 boys, 12 girls, age = 8.7
FSIQ = 108.4, VIQ = 100.1, PIQ = 117.8
Highest subtest—Coding, Lowest subtest—Vocabulary
Performance IQ greater than Verbal IQ ($P \leq .05$)

II. 24 boys, 9 girls, age = 9.2
FSIQ = 94.5, VIQ = 101.5, PIQ = 88.6
Highest subtest—Comprehension, Lowest—Block Design
Verbal IQ greater than Performance IQ ($P \leq .05$)

III. 25 boys, 11 girls, age = 10.1
FSIQ = 95.5, VIQ = 87.7, PIQ = 105.5
Highest subtest—Object Assembly, Lowest—Information
Performance IQ greater than Verbal IQ ($P \leq .05$)

IV. 9 boys, 4 girls (smallest cluster N = 13), age = 8.1 (youngest)
FSIQ = 97.1, VIQ = 99.0, PIQ = 94.5
Highest subtest—Comprehension, Lowest—Vocabulary
Profile contains an abnormal amount of scatter when compared to WISC-R standardization sample.

V. 17 boys, 23 girls (atypical ratio), age = 9.7
FSIQ = 81.4, VIQ = 78.1, PIQ = 87.7
Highest subtest—Comparison, Lowest—Information
Group has lowest full scale IQ.
Performance IQ greater than Verbal IQ ($P \leq .05$).
Exhibits Bannatyne profile.

VI. 26 boys, 6 girls, age = 10.4 (oldest)
FSIQ = 111.1, VIQ = 107.7, PIQ = 112.9
Highest subtest—Picture Arrangement, Lowest—Coding
Group has the highest full scale IQ.
Exhibits Bannatyne profile.

VII. 25 boys, 9 girls, age = 9.9
FSIQ = 91.4, VIQ = 90.7, PIQ = 93.6
Highest subtest—Picture Arrangement, Lowest—Arithmetic
Supports Swartz's ACID profile somewhat.

VIII. 21 boys, 4 girls, age = 10.1
FSIQ = 96.4, VIQ = 96.7, PIQ = 97.1
Highest subtest—Similarities, Lowest—Vocabulary
Exhibits Bannatyne profile though difference between scores is quite small.

exhibit an abnormal amount of scatter, when "abnormal" is determined by comparison to score scatter within the *WISC-R* standardization sample. Finally, a rather discouraging finding is that not one of the eight performance profiles discovered in this study supports any of the five

> performance profiles suggested by Vance, Wallbrown, and Blaha (1978). Only cluster IV is similar to a Vance et al. Profile—perceptual organization. A low vocabulary score, however, is not included in the perceptual-organization profile and cluster IV is characterized by an extremely low vocabulary score. (Bates, 1981, pp. 85–87)

In another study of within-group differences, Sinclair and Kheifets (1982) applied cluster analysis to identify subgroups among 42 nine- to eleven-year-old boys with school learning problems. One group was determined to be mildly handicapped and described as having only minimal impairments in cognitive functioning and mild to moderate problems in academic achievement. Cluster II consisted of children with moderate to severe problems in cognitive functioning, sensorimotor/perceptual functioning, and academic achievement. Cluster III consisted of children who showed significant differences in cognitive functioning.

DEVELOPING A FRAMEWORK

An alternative aimed at addressing the issues we have discussed is to consider mathematics programming for the learning disabled within some general framework. Figure 1–3 illustrates one such framework.

Content

The initial step is to designate the set of content that will be presented to the individual and to determine the age or level most appropriate for the presentation of this content. To do so, we need the following:

1. An assessment of a wide range of mathematics content among the learning disabled across age levels. It is necessary to determine what children can do at certain developmental levels (e.g., what do five-year-olds know and do) and when certain skills and concepts are mastered (e.g., when do LD children do $\frac{1}{2} \times \frac{1}{4}$).
2. Continuous exposure to a wide range of content to prevent gaps in one or more areas (e.g., geometry) or in one or more concepts (e.g., $2 + 3 = 5$ and $3 + 2 = 5$ as indicators of commutativity) so that the concept will be properly applied at a later stage (e.g., $\frac{1}{2} + \frac{1}{4} = \frac{3}{4}$ and $\frac{1}{4} + \frac{1}{2} = \frac{3}{4}$).
3. Continued integration of concepts, skills, problem solving, and applications at all levels.
4. Prevention of habituation of error patterns or erroneous concepts.

Figure 1–3 Framework for Mathematics Programming

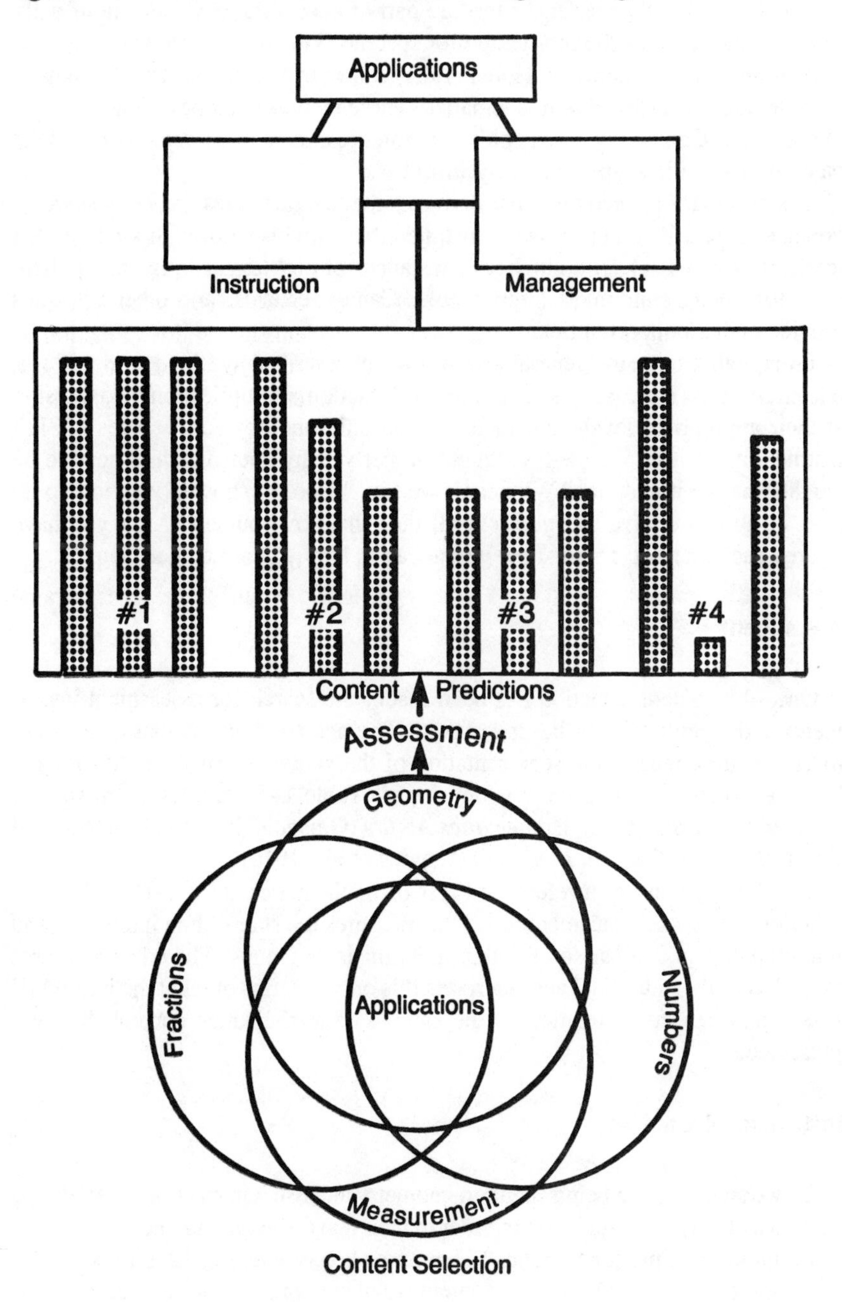

With a cognitive-reasoning approach to mathematics with the learning disabled the applications and experiences that are part of the natural environment of daily living would serve as the motivation for specific skill and concept learning.

A summary of literature by Hollis (1981) shows that very young children have considerable background in mathematics. These include the topological concepts of open-closed among two-year-olds, congruence among three-year-olds, and the naming of specific shapes by five-year-olds.

It is necessary to recognize the diversity that characterizes the knowledge of young children. Kurtz (1978) demonstrated this with his work with kindergarten teachers (Table 1–4). The number of instances of problem solving and applications for young children is innumerable. Games, errands, and other activities provide endless opportunities.

To capitalize upon the generalization strategy described by Lloyd et al. (1981), one might stress the concepts of addition, subtraction, multiplication, and division at their appropriate levels and then provide only one or two instances of skill instruction. That is, if 3 + 2 is taught properly, why does the child need to be taught other combinations? Why can't we say, "Look, I showed you how to do 3 + 2, now you have to figure out all the others by yourself. After you have figured each one out, I will show you a way to help you remember them."

Assessment

Once the content selection has been made, the search for assessment instruments and techniques can be undertaken. The one condition is that assessment must contain a reasonable representation of the content of mathematics for the levels at which the assessment is developed. Table 1–5 provides a comparison between the *Test of Early Mathematics Ability* (Ginsburg & Baroody, 1983) and the *Mathematics Concept Inventory* (Cawley et al., 1976).

The TEMA is a norm-referenced test of mathematics ability. The MCI is a criterion-referenced placement test. One measures the status of an individual and interprets this in comparison to other individuals or groups. The other measures the status of the individual and interprets this only in terms of what the individual knows and can do in relation to the sample of mathematics content that was presented.

Instructional Choices

Do we take the time being spent on geometry and use it in fractions? Or, do we cut down on physical education in order to maintain the geometry and still provide some additional time for fractions? These are choices and they have to be made. Given the fact that we know what content we plan to present throughout the 12 or

Table 1–4 Competencies

Competency	Percent of Children Reaching Competency 0–25	26–50	51–76	76–100
Clearly Kindergarten Competencies				
Rationally count to 20	2	3	15	82*
Recognition of numerals to 10	0	1	7	92
Identify sets of 0, 1, 2, 3, 4, and 5	0	3	8	89
Identify sets of 6, 7, 8, 9, and 10	3	4	17	76
Write numerals 0–5 in order	2	2	4	92
Write numerals 6–10 in order	1	4	9	86
Identification of circle, square, rectangle, and triangle	0	2	4	94
Identification of a circle, its inside and outside	2	5	12	81
Questionable Kindergarten Competencies				
Write numerals in order to 20	13	17	37	33**
Identify sets for one more and one less	5	10	39	46
Join 2 sets to form sums to 5	7	16	30	47
Join 2 sets to form sums to 10	17	21	30	32
Separate 2 sets to form differences from 5	23	23	22	32
Add numerals to sums of 5	12	22	26	40
Tell time on the hour	15	20	31	34
Locate a day of the month on the calendar	9	18	33	40
Identify penny, nickel, and dime	6	18	34	42
Clearly Not Kindergarten Competencies				
Write numerals as sets of tens and ones	66†	15	12	7
Rationally count to 100	25	22	37	16
Recognition of numerals to 50	19	31	37	13
Identify halves, fourths, and thirds	40	37	18	5

Table 1–4 continued

Competency	*Percent of Children Reaching Competency*			
	0–25	*26–50*	*51–76*	*76–100*
Clearly Not Kindergarten Competencies				
Count by twos to 20	57	24	14	5
Separate 2 sets to form differences from 10	34	24	24	18
Add numerals to sums of 10	27	22	29	22
Subtract numerals to form differences from 5	35	25	17	23
Subtract numerals to form differences from 10	48	23	21	8
Read a thermometer	49	27	20	4
Identify odd numbers to 20	64	17	14	5

*This is read: 82 percent of the teachers reported that at the end of the school year, 76–100 percent of their children could rationally count to 20. This percentage added to the next lower figure indicates that almost all children do well on this competency.

**This is read: 33 percent of the teachers reported that at the end of the school year 76–100 percent of their children could write numerals in order to 20.

†This is read: 66 percent of the teachers reported that at the end of the school year only 0–25 percent of their children could write numerals as sets of tens and ones.

Source: Kurtz, V.R. (1978). Kindergarten mathematics—a survey. *Arithmetic Teacher, 25,* 51–53. Used by permission.

14 years of school and the approximate level when different individuals will be ready for the content, we can initiate choices relative to instructional methodology and individual or group management.

Management refers to the procedures used to maintain attention or on-task behavior, increase or decrease rates of performance, or to derive secondary benefits from the activity, such as learning to work independently.

Instruction is the procedure through which concepts and skills are taught within the framework of some management orientation. The choice to model for a child is an instructional choice. The decision to use manipulatives or to use paper and pencil activities are instructional choices. The use of a question such as "How many ways can you write five?" is an instructional decision that is different from one such as "Listen. Repeat after me, two plus three is five."

Table 1–5 Content Comparisons between *Test of Early Mathematics Ability* and *Mathematics Concept Inventory* for Similar Ages

	TEMA			MCI		
Age Range	*Strand*	*Area*	*Concept*	*Strand*	*Area*	*Concept*
4–0 to 4–11	Numbers	Counting	Count after me Count out loud Enumeration			
		Cardinal Property	Concept of More			
		Addition	Adding Objects			
5–0 to 6–11				Geometry	Topology	Open-Closed Order Constancy Inside-Outside-On Betweenness
					Shape	Two-dimensional Three dimensional
					Relations	Congruence Similarity Symmetry
				Sets	Classification	Grouping Empty Set
					Correspondence	One-to-One
					Operations	Set Union Set Intersection

Table 1–5 continued

	TEMA			MCI		
Age Range	*Strand*	*Area*	*Concept*	*Strand*	*Area*	*Concept*
5–0 to 6–11 continued				Patterns	Perceptual Characteristics	Original Learning Copy
						Original Learning Extend
						Intradimensional shift Copy
					Language	Original Learning Copy
						Original Learning Extend
			Count Back From 10			
			Produce 19 Counting Out Loud 41:			
			Count By Tens			
			Count After Me: 11			
			Enumeration: Large			
			Count After Me: 111			

5–0 to 6–11 continued	Cardinal Property	Reading Numerals: Teens		Cardinal Property	Fewer Number/Greater Number
		Writing Two-Digit Numerals			Oneness
					Cardinal Numbers 1–10
					Zero
		Mental Number Line			Reading and Writing Number Names
		Reading Numerals: Two Digits			Different Names for Same Number
					Cardinal Numbers 1–19
					Skip Counting
				Ordinal Numbers	First and Last
					Order and Position
	Addition	Mental Addition: Pennies		Addition	Joining Sets
					Signs of Operation
					One-Digit Numbers
					Identity Element
				Subtraction	One-Digit Numbers
					Identity Element
				Place Value	Grouping By Tens
				Whole Numbers	Even, Odd
			Measurement	Height	High/Low
					Tall/Short
				Temperature	Event

Table 1–5 continued

	TEMA			MCI		
Age Range	*Strand*	*Area*	*Concept*	*Strand*	*Area*	*Concept*
5–0 to 6–11 continued					Time	Sequencing Days of Week AM/PM Telling Time Calendar
					Speed	Rate
				Fractions	Part/Whole Relations	Discrete Parts
						Blended Parts
						Replaceable Parts Equal Parts of Whole Part-Size

Applications

At every level, and as an ultimate program product, applications are the goal. One set of applications will be used to further the growth of the individual in mathematics. Another set of applications will be social and utilitarian. We should not confuse one with the other. For example (Cawley, 1981), 3 apples + 6 apples + 4 pears is a utilitarian activity at one level. If the process is overgeneralized when $3x + 6x + 4y$ is given, the mathematical appropriateness is distorted by the utilitarian activity. "How many halves in ten?" is a mathematics item. "How many half-inch pieces of wood can you get from a board ten inches long?" is a utilitarian item and the response is likely to be different (probably 18 instead of the expected 20), because of wastage in the process of sawing).

Within the schemata of applications is the domain of problem solving and, in particular, word problems. Word problems (Cawley, 1982) are problems in which the processing of the information contained in the words is what makes the problem a problem and not a computational activity surrounded by words. If, in fact, the use of the "problem" is to provide practice in computation, why not simply practice the computation? If, on the other hand, the purpose of the word problem is to give the child something to reason about, practices such as giving cue words will have to be eliminated.

There is very little research in the area of problem solving among the learning disabled. What there is in word problems focuses upon syntactical and semantic features of the problems. There is not enough information available to draw conclusions or to make inferences. The time has come to mount a concerted effort in word problems.

Ballew and Cunningham (1982) provide an interesting perspective from which an approach to word problems could be initiated. These authors set up a profile that consists of a Computation Score (the highest level at which the student could correctly compute 75 percent of the problems), a Problem Interpretation Score (the highest level at which the student could correctly set up 75 percent of the problems when they were read to the student), a Reading-Problem Interpretation Score (the highest level at which the student could set up 75 percent of the problems when they were read by the student), and a Reading-Problem Solving Score (the highest level at which 75 percent of the problems were solved correctly). Their data showed considerable variation in the type of difficulty manifested by different students. Only 19 of 217 complete profiles showed all four scores to be the same, a fact which indicates that the students were *rarely* equally good or equally poor in solving word problems. Only 12 percent of the students could read and set up problems correctly at a level higher than they could compute, whereas 60 percent could compute correctly at a higher level than they could read and set up the problems. A multifaceted approach similar to that used by Ballew and Cunningham would seem to be in order for the field of learning disability.

CONCLUSION

Our alternatives, both from the teaching and research perspectives, are numerous. We can try new curricula, new algorithms, new diagnostics, and many other things. Whatever we do, the following need to be kept in mind:

- *Comprehensiveness.* Learning-disabled children attend school for some 12 to 14 years. Program development needs to recognize this reality.
- *Individualism.* Regardless of what may be stated or written about groups of learning-disabled persons, each individual is unique.
- *Correction.* The performance of each individual must be examined for consistency. Determine the extent to which the same individual performs the same items the same way under the same condition. Change one component at a time. If for example, a child subtracts going left to right and makes errors with the number facts, do not change the left-to-right approach. Correct the errors in the number facts. Excessive correction can cause as many difficulties as neglect.
- *Alternatives.* If one approach does not work, try another.
- *Intervention.* Do not permit the child to habituate inappropriate concepts or algorithms. Intervene early. It is much more difficult to convince a child that his or her way of doing something is not correct than it is to teach the task properly in the first place.

These guidelines offer a beginning. The remainder of this book will expand upon them.

REFERENCES

Ballew, H., & Cunningham, J. (1982). Diagnosing strengths and weaknesses of sixth-grade students in solving word problems. *Journal for Research in Mathematics Education, 13,* 202–210.

Bates, H.M. (1981). *An application of cluster analysis to identify homogeneous groups of learning disabled students on the basis of WISC-R subtest scale score patterns.* Unpublished doctoral dissertation, University of Connecticut, Storrs.

Carpenter, T., Corbitt, M., Kepner, H., Lindquist, M., & Reys, R. (1980). Results and implications of the second NAEP mathematics assessment: Elementary school. *Arithmetic Teacher, 27,* 10–47.

Cawley, J.F. (1981). Commentary. *Topics in Learning and Learning Disabilities, 1,* 89–95.

Cawley, J.F. (1982). *Word problem activities for children with handicaps to learning and achievement: An integrated approach.* Paper presented at the meeting of the Research Council for Diagnostic and Prescriptive Mathematics, Buffalo, N.Y.

Cawley, J., Fitzmaurice, A.M., Goodstein, H.A., Lepore, A., Althaus, V., & Sedlak, R. (1976). Mathematics Concept Inventory, Levels I and II. *Project MATH.* Tulsa, Okla.: Educational Progress Corp.

Cawley, J.F., Fitzmaurice, A.M., Shaw, R.A., Kahn, H., & Bates, H.M. (1979). LD youth and mathematics: A review of characteristics. *Learning Disability Quarterly, 1,* 29–44.

Clements, S.D., & Peters, J.E. (1962). Minimal brain dysfunctions in the school-age child. *Archives of General Psychiatry, 6,* 185–197.

Fleischner, J., Garnett, K., & Shepherd, M. (1980). *Proficiency in arithmetic basic fact computation of learning disabled and non-disabled children* (Tech. Rep. No. 9). New York: Teachers College, Columbia University, Research Institute for the Study of Learning Disabilities.

Ginsburg, H.P., & Baroody, A.J. (1983). *The Test of Early Mathematics Ability.* Austin, Tex.: Pro-Ed.

Glennon, V., & Cruickshank, W. (1981). Teaching mathematics to children and youth with perceptual and cognitive processing deficits. In V. Glennon (Ed.), *The Mathematical Education of Exceptional Children and Youth.* Reston, Va.: The National Council of Teachers of Mathematics.

Greenes, C. (1979). The learning disabled child in mathematics. *Focus on Learning Problems in Mathematics, 1,* 5–11.

Hammill, D., Leigh, J., McNutt, G., & Larson, S. (1981). A new definition of learning disabilities. *Learning Disability Quarterly, 4,* 336–343.

Harber, J. (1981). Learning disability research: How far have we progressed? *Learning Disability Quarterly, 4,* 372–382.

Hollis, L.Y. (1981). Mathematics concepts for the very young. *Arithmetic Teacher, 29,* 24–27.

Johnson, S. (1979). *Arithmetic and Learning Disabilities.* Boston: Allyn & Bacon.

Kavale, K., & Nye, C. (1981). Identification criteria for learning disabilities: A survey of research literature. *Learning Disability Quarterly, 4,* 383–388.

Kurtz, V.R. (1978). Kindergarten mathematics—a survey. *Arithmetic Teacher, 25,* 51–53.

Lepore, A. (1979). A comparison of computational errors between educable mentally handicapped and learning disabled children. *Focus on Learning Problems in Mathematics, 1,* 12–33.

Lloyd, J., Saltzman, N., & Kauffman, J. (1981). Predictable generalization in academic learning as a result of preskills and strategy training. *Learning Disability Quarterly, 4,* 203–216.

Loring, R. (1981). The cognitive nature of learning disabilities in mathematics. *Dissertation Abstract International, 4,* 4710–B.

Meltzer, L. (1978). Abstract reasoning in a specific group of perceptually impaired children: Namely the learning disabled. *Journal of Genetic Psychology, 132,* 185–195.

Myers, P., & Hammill, D. (1982). *Learning Disabilities: Basic Concepts, Assessment Practices and Instructional Strategies.* Austin, Tex.: Pro-Ed.

Olson, J., & Mealor, D. (1981). Learning disabilities identification: Do researchers have the answer? *Learning Disability Quarterly, 4,* 389–392.

Pieper, E., & Deshler, D. (1980). *A comparison of learning disabled adolescents with specific arithmetic and reading disabilities* (Research report No. 27). University of Kansas, Research Institute for Research in Learning Disabilities.

Reisman, F. (1982). *A Guide to the Diagnostic Teaching of Arithmetic.* Columbus, Ohio: Charles E. Merrill.

Sinclair, E., & Kheifets, L. (1982). Use of clustering techniques in deriving psychoeducational profiles. *Contemporary Educational Psychology, 7,* 81–89.

Swanson, L. (1982). Conceptual process as a function of age and enforced attention in learning disabled children: Evidence for deficient rule learning. *Contemporary Educational Psychology, 7,* 152–160.

Swartz, G.A. (1974). *The Language-Learning System.* New York: Simon & Schuster.

Tabachnick, B. (1979). Test scatter on the WISC-R. *Journal of Learning Disabilities, 12*(9), 60–62.

U.S. Congress, P.L. 94-142, Education for All Handicapped Children Act, 1975.

Vance, H.B., Wallbrown, F.H., & Blaha, J. (1978). Determining WISC-R profiles for reading disabled children. *Journal of Learning Disabilities, 11*, 657–661.

Webster, R. (1981). *Learning efficiency test*. Novato, Calif.: Academic Therapy Publications.

Chapter 2

Curriculum and Instruction: An Examination of Models in Special and Regular Education

Colleen S. Blankenship

Educators have continually tried to answer the following questions:

1. *What* mathematics should students learn?
2. *How* should they be instructed?

Mathematics educators have generally regarded determining "what" to teach as the easier of the two questions to answer (Glennon, 1965).

"What" mathematics to teach normally achieving students has changed over the years according to the prevailing wisdom of the time. Prior to the turn of the century, mathematics instruction was viewed as a vehicle for training the mind. By the 1920s the focus changed to teaching students practical computational skills in order to prepare them for the world of work. From the mid-1950s to 1975, the "new math" emphasized understanding the logical structure of mathematics (Glennon, 1975). In recent years the pendulum has swung "back-to-basics."

While mathematics educators have concerned themselves with determining "what" to teach, psychologists and educational psychologists have focused on "how" to teach. Despite years of research, no one instructional method has been found to be clearly superior to another (Suydam & Osborne, 1977). Today, we have an extensive array of instructional approaches for teaching mathematics, including expository, discovery-learning, math labs, programmed instruction, computer-assisted instruction, and direct instruction.

At present, there is a lack of consensus concerning "what" mathematics to teach normally achieving students and "how" best to instruct them (Trafton, 1981). Determining "what" mathematics to teach normally achieving students will perhaps never be fully resolved, nor should it be. The curriculum must necessarily reflect the changing needs of society and the prevailing view of the purposes for studying mathematics. Similarly, "how" to teach mathematics will perhaps never be agreed upon, nor should it be. By limiting ourselves to current teaching methods we run the risk of failing to adopt promising new approaches validated by research or to incorporate technological advances into teaching.

While considerable time has been spent determining "what" and "how" to teach mathematics to normally achieving students, educators have rarely focused on trying to answer these questions for handicapped students. In the field of special education in general, and in the area of learning disabilities in particular, "what" mathematics to teach has received even less emphasis than "how" to teach. Because learning disabilities has become almost synonymous with reading difficulties, little attention has been paid to the mathematical difficulties of learning-disabled students (Reid & Hresko, 1981b). While special instructional methods abound, applications of these strategies to improve the *mathematical* abilities of learning-disabled students are almost nonexistent.

Until quite recently, regular educators have not had to confront the problems of providing instruction in mathematics to students classified as learning disabled. Since the advent of mainstreaming, many students with learning problems have been allowed to remain in or return to the regular classroom, and, as a result, regular educators are assuming a greater responsibility for their education. Many classroom teachers feel unprepared to teach students with learning problems and have expressed concern that they "often fail to recognize the depth of these students' problems and underestimate the kind of help they need" (Trafton, 1981, p. 20).

While it is natural for regular educators to rely on special educators for assistance in determining "what" mathematics to teach learning-disabled students and "how" best to instruct them, our knowledge in this area has been described as virtually "near zero" (Cawley, 1977, p. 33).

Perhaps a reasonable way to begin to answer the "what" and the "how" of providing mathematics instruction to the learning disabled is to examine current practices in regular and special education. It will be useful to evaluate the standard K–12 mathematics curriculum in terms of the difficulties it presents to students with learning disabilities and to assess the appropriateness of current instructional practices for teaching the learning disabled. The applicability of special mathematics curricula and specialized instructional approaches for teaching mathematics to the learning disabled also will be examined.

THE K–12 MATHEMATICS CURRICULUM

According to the most recent figures from the U.S. Department of Education (1980), approximately 80 percent of the students classified as learning disabled receive all or at least part of their instruction in the regular classroom. The number of learning-disabled students who receive mathematics instruction in regular rather than resource rooms is not known. On the basis of the following facts, one may assume, however, that it is significantly high.

1. It is more common for teachers to refer students for reading rather than for mathematics difficulties (McLeod & Armstrong, 1982; Reid & Hresko, 1981b).
2. Resource teachers report spending more time teaching reading than mathematics (Houck & Given, 1981).
3. Classroom teachers indicate that there is a lack of curriculum materials to teach mathematics to slower students (Denny, 1978).

Curriculum Models

The school mathematics curriculum has been described by Robitaille and Dirks (1980) "as being an outgrowth of the nature of the subject itself, adapted, filtered, and restructured by the curriculum development process" (p. 21). A schematic representation of the curriculum development process as conceptualized by Robitaille and Dirks is shown in Figure 2–1. It is a process influenced by sociological, psychological, pedagogical, and technological factors.

- *Sociological Factors*. The values of society are reflected in the curriculum. For example, society's demand for accountability has exerted pressure to emphasize basic mathematical skills.
- *Psychological Factors*. Beliefs about ways students learn and what they are capable of learning at different ages influence both "what" mathematics students learn and "how" they are taught. The theories of Thorndike (1924), Bruner (1963), Piaget (1973), and Gagné and Briggs (1974) have all influenced the nature of mathematics education.
- *Pedagogical Factors*. The methods and materials used by teachers as well as their knowledge of mathematics influence the curriculum and hence the content attained by students.
- *Technological Factors*. Technological innovations such as hand-held calculators and microcomputers have implications for the content of the curriculum, the emphasis to be accorded to various topics, and the way in which instruction is provided.

It is possible to conceive of several different curriculum models based on one's view of the nature of mathematics as influenced by sociological, psychological, pedagogical, and technological factors.

The three most common models today are the pure mathematics model, the applied mathematics model, and the basic mathematics model (Robitaille &

Figure 2–1 Development of a Mathematics Curriculum Model

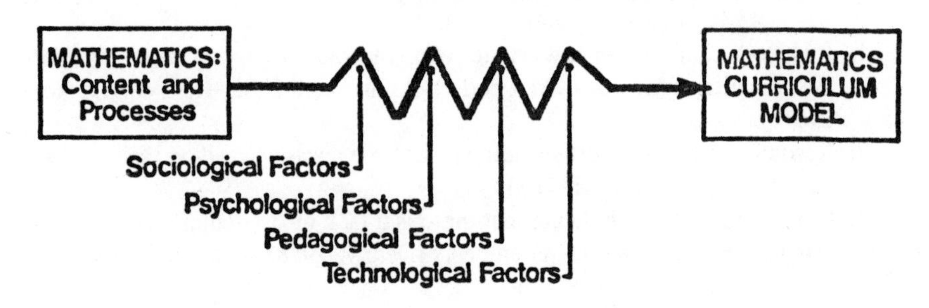

Source: Robitaille, D.F., and Dirks, M.F. (1982). Models for the mathematics curriculum. *For the Learning of Mathematics, 2*(3). Used by permission.

Dirks, 1980). In the pure mathematics model the logical structure of mathematics is emphasized. Curriculum during the "new math" era was heavily influenced by the structure of the discipline approach of the pure mathematics model. The applied mathematics model stresses applications of mathematics to other disciplines. The basic mathematics model stresses the use of mathematics in daily living. In commenting on these models, Robitaille and Dirks (1980) stated:

> These three models are seldom if ever found in a pure state. Thus, the mathematics curriculum for a given place and for a given grade level might advocate placing greater emphasis on one model than on the others, but it is extremely unlikely that a curriculum would be constructed on the basis of one of these models to the total exclusion of the other two. (p. 23)

In describing what the nature of the curriculum should be, mathematics educators recommend a balance among concepts, skills, and applications (Trafton, 1981), areas reflected in the pure, applied, and basic mathematics models. Most basal mathematics textbooks attempt to address this concern.

Assessing the Current Curriculum

Much of what is known concerning current practices in curriculum and instruction in mathematics, science, and social studies stems from three studies commissioned by the National Science Foundation (NSF). In each subject area the following types of studies were conducted:

1. A literature review on curriculum, instruction, and evaluation for the 20-year period beginning with 1955. In mathematics the literature review was written by Suydam and Osborne (1977).
2. A national survey of teachers, supervisory personnel, parents, and students in grades K–12 concerning their views on curriculum and instruction in science, mathematics, and social studies. This was directed by Weiss (1978).
3. A collection of case studies consisting of observations in classrooms within selected school districts and interviews with teachers, administrators, supervisory personnel, and students. This was undertaken by Stake, Easley, and their collaborators (1978).

In addition to these three sources, an earlier study conducted for the National Advisory Committee on Mathematical Education (NACOME, 1975) provides valuable insight into the content and nature of mathematics instruction in this country.

One of the most common findings in the NSF studies was that "The textbook is the primary determinant of mathematical curricula throughout schools in this country" (Suydam & Osborne, 1977, p. 84). This conclusion was supported by Weiss (1978), who found that 94 percent of the teachers in her survey reported using one or more mathematics textbooks. Nearly two-thirds of the teachers in Weiss's study indicated that they based instruction on a *single* textbook.

While regular educators may rely on a single mathematics textbook, this practice appears to be less common among LD teachers. In a survey of 114 teachers employed in grade level 6 or above (McLeod & Armstrong, 1982), the majority of LD teachers reported basing mathematics instruction on other commercial materials or adapting the text used in LD students' regular classrooms.

The most popular texts at the elementary level have been described by Fey (1980) as traditional programs that "include arithmetic of whole numbers, common fractions and decimals, measurement, geometry, descriptive statistics, and applications of these topics" (p. 34). While textbooks do include work on geometry, statistics, and measurement, the extent to which teachers stress these topics has been questioned. The results of a survey reported by NACOME (1975) suggest that "new math" topics such as geometry and statistics "have made little headway against the traditional domination by arithmetic computation" (p. 73). After interviews with numerous K–12 mathematics teachers, Denny (1978) concluded that "nine teachers in ten follow the mathematics textbook, 'sorta'" (p. 32). A similar conclusion was reached by Suydam and Osborne (1977), who noted that while teachers try to "cover the material" they tend to pick and choose among topics, ignoring those they feel are unimportant. The following comments made by teachers interviewed by Denny (1978) suggest that what's "out" is geometry and conceptual mathematics and what's "in" is learning the 100 basic facts in each of the four arithmetic operations (p. 1–18).

> This book has too much esoteric garbage in it. It is simply too hard. The geometry is silly (to try and teach) even for our best third graders. So we all skip it. (p. 1–33)

> Modern math? I dislike it. Too many ways befuddle [a] young child. I skip what I think is useless and use what I think is pertinent. (p. 1–31)

> You might as well forget about teaching conceptual mathematics to 75% of the children in elementary school. The upper level children like it. The rest are not only bored—they hate it. (p. 1–34)

> I dislike our book, not enough drill, it's modern math. We adopted a new book. I don't know its title, but it has more drill, more basics, and I'll like it. (p. 1–31)

If the focus of the curriculum for average youngsters is on computational skills, one wonders what the emphasis is for lower-achieving students. Although his case study did not specifically address practices with learning-disabled students, Denny noted that the lowest-level group in mathematics often included special education students. The "what" of mathematics for slower students appears to be the proverbial watered-down curriculum. The following comments serve to characterize the nature of the curriculum provided to slower students and to convey the frustration many regular educators experience in teaching them.

- *Elementary Teacher:* We just do not have curriculum for the child who has difficulty in arithmetic. What we do is slow down, take smaller bites, do it more often and pray. (Denny, 1978, p. 1–32)
- *Junior High Teacher:* Level three mathematics is often taught a grade or two, or three, below the child's current class assignment. It consists often of practice on the basic facts in multiplication or division with some attention to fractions. Story problems of the garden variety are used when reading skills permit. (Denny, 1978, pp. 1–66, 1–67)
- *High School Teacher:* That kind of student I just don't know how to work with. Nor do I have the time. The quarter marches on, you know. So we give them more of the same, perhaps slower, perhaps louder; then when that doesn't work we just threaten them . . . and finally ignore them. (Denny, 1978, p. 1–78)

It is no wonder that some students view mathematics as "learning over and over what you already know and keep forgetting" (Denny, 1978, p. 1–72).

Mathematics educators (Trafton, 1981), special educators (Cawley, 1977), and teachers (Denny, 1978) have expressed dissatisfaction with the standard curriculum for teaching students with learning problems. Some of the specific criticisms of basal mathematics texts noted by teachers in Denny's (1978) study were as follows:

1. The reading levels were perceived as being too high. This observation is supported by Suydam and Osborne (1977), who noted that the readability level of mathematics texts has been a concern for the past ten years.
2. Sequencing was poor, frequently skipping from one concept to the next.
3. The number of problems provided was often insufficient to ensure mastery.
4. Too few applications problems were provided.
5. Formats varied every few pages, moving from left to right, top to bottom, and back again.
6. More than one method was presented to teach a skill such as regrouping in addition. After teaching Method A, teachers were directed by their manuals to switch to Method B. As this practice confused students, teachers reported that they skipped demonstrating Method A.
7. Texts assumed students possessed prerequisite skills that they had in fact not yet acquired.

In order to circumvent many of the difficulties inherent in basal mathematics textbooks, some publishers have developed alternate texts. One example is the *Mathematics for Individual Achievement* series (Denholm, Hankins, Herrick, & Vojtka, 1977). In this series the authors have endeavored to reduce the role of reading, to simplify methods of presentation, to incorporate multisensory learning, and to make use of a laboratory approach. Data on the effectiveness of such alternate texts with the learning disabled is lacking. However, the existence of such texts further underscores the difficulties that students with learning problems often experience with standard curriculum materials.

Implications for Teaching the Learning Disabled

When the majority of teachers report that they use a single text to teach mathematics, one can anticipate the kinds of difficulties students with learning problems will experience. First, unless a student is placed into an appropriate level, we cannot expect him or her to profit from instruction. The extent to which teachers routinely assess students' skills prior to making curricular decisions is a critical factor. Skager (1969) found that teachers often targeted objectives for low-achieving seventh graders that they had already mastered. Similarly, in a pilot study conducted by the Educational Products Information Exchange (EPIE, 1976), it was determined that students on the first day of school achieved an average score of 64 percent correct on items taken from the curriculum that was to be used during the current year.

Assessment prior to instruction is important for all students but it is crucial for those with learning difficulties. While textbook publishers have begun to develop placement tests, the adequacy of many of these devices is questionable as they frequently assess performance on only one occasion, rarely provide a representative sample of items, and fail to provide information that is diagnostically useful.

An alternative to standardized tests is curriculum-based assessments, teacher-made devices designed to measure student performance on a series of sequentially arranged objectives extracted from the curriculum to be used for instruction (Blankenship & Lilly, 1981). This assessment strategy allows teachers to determine "what" skills students have and have not mastered prior to instruction. Using this method, one could at least ensure that students were placed appropriately.

Second, even though a teacher has assessed a pupil's performance prior to placing him or her into a basal text, other modifications must often be made to meet the needs of students with learning problems. As basal texts frequently do not contain sufficient examples and problems, teachers must supplement the curriculum to ensure mastery of skills.

Third, a number of students experience reading difficulties that interfere with their ability to read directions or to compute application problems. Therefore, adjustments must be made to compensate for students' reading problems if they are placed into basal mathematics texts.

Fourth, mathematics instruction commonly proceeds "unit" by "unit." Because so many mathematical skills are hierarchical in nature, unless the student has mastered the prerequisite skills he or she will experience difficulty on more advanced skills. By locking students into such a "fixed frequency system" (Cawley, 1978), we ensure that some students will not master necessary concepts and skills.

Fifth, by repeatedly presenting students with material that is too difficult we almost guarantee that they will come to view mathematics as nothing more than an endless series of problems for which they seldom have the correct answers.

INSTRUCTIONAL PRACTICES IN REGULAR EDUCATION

In recent years teachers have been encouraged to use a variety of instructional approaches, including discovery learning, mathematics labs, programmed instruction, and multimedia instruction (Fey, 1980). The extent to which these instructional innovations have been implemented has been questioned (NACOME,

1975; Suydam & Osborne, 1977; Weiss, 1978). For that reason, we will examine instructional practices that are currently being implemented in regular classrooms and assess their appropriateness in teaching learning-disabled students.

Assessing Current Instructional Practices

Weiss (1978) analyzed the frequency with which teachers reported using various instructional approaches in mathematics education. Table 2–1 presents a summary of Weiss' findings. It appears that the predominant instructional approach consists of lecture/discussion, teacher demonstration, chalkboard work, and the completion of individual assignments. This view of mathematics instruction was supported by Welch (1978), who concluded the following based on his case study observations:

> In all mathematics classes I visited, the sequence of activities was the same. First, answers were given for the previous day's assignment. The more difficult problems were worked by the teacher or a student at the chalkboard. A brief explanation, sometimes none at all, was given of the new material, and problems were assigned for the next day. The remainder of the class was devoted to working on the homework while the teacher moved about the room answering questions. The most noticeable thing about mathematics class was the repetition of this routine. (pp. 5–6)

If the primary instructional approach for normally achieving students was the "tell and drill" method, one wonders whether teachers individualized instruction for slower students. If we interpret Weiss's findings in light of the instructional methods used, it appears that those that lend themselves to individualizing instruction were underutilized. For example, according to the data shown in Table 2–1, 75 percent of the teachers had never used programmed instruction. Using programmed texts, lower-achieving students could at least progress through the materials at their own rate. Another instructional approach that allows for individualization is computer-assisted instruction (CAI). Over 90 percent of the teachers indicated that they had never used CAI. Now that microcomputers are appearing with increasing frequency in schools and mathematics educators are calling for "computer literacy," computers may provide an efficient method for individualizing instruction. A third instructional approach that allows for individualization is the use of contracts where students agree to complete individual packets of materials stressing mathematical skills. Many fourth and fifth grade teachers in Denny's (1978) study reported that they tried to individualize instruction for both low and high achievers by using contracts. In assessing the utility of

Table 2–1 Frequency of Use of Various Techniques in Mathematics Classes

	Percent of Classes					
Technique	*Never*	*Less than once a month*	*At least once a month*	*At least once a week*	*Just about daily*	*Missing*
Lecture	23	4	3	21	46	4
Discussion	5	2	3	16	71	2
Student reports or projects	46	28	15	5	4	4
Library work	74	16	2	4	1	4
Students working at chalkboard	5	8	13	36	36	2
Individual assignments	9	7	5	17	59	3
Students use hands-on manipulative or laboratory materials	19	23	16	24	14	4
Televised instruction	87	5	2	4	0	2
Programmed instruction	75	7	6	3	4	5
Computer-assisted instruction	91	3	2	2	1	2
Tests or quizzes	5	5	26	56	6	2
Contracts	78	7	5	3	4	3
Simulations (role-play, debates, panels)	81	8	5	4	1	2
Field trips, excursions	78	19	1	0	0	2
Guest speakers	86	10	1	0	0	2
Teacher demonstrations	11	9	12	28	36	4

Source: Adapted from Weiss, I. (1978). *Report of the 1977 national survey of science, mathematics, and social studies education.* Research Triangle Park, N.C.: Research Triangle Institute. Used by permission.

this approach, however, teachers noted that students were sometimes more interested in finishing the packets than in understanding the contents.

Noting the infrequent use of methods to individualize instruction, one must conclude that "teachers frequently do not differentiate instruction" for students with learning problems (Suydam & Osborne, 1977, p. 50).

Implications for Teaching the Learning Disabled

Presumably, instruction for learning-disabled students in regular classrooms is also characterized by whole group instruction followed by practice on individual assignments. Teachers in Denny's (1978) study reported a greater reliance on rote methods in teaching slower students. Given the difficulty those teachers reported in maintaining student attention using the "tell and drill" method, it is likely that many learning-disabled students fail to fully profit from instruction.

Achievement is correlated with the amount of time students spend actively engaged in instruction (Stevens & Rosenshine, 1981). Elementary teachers in Weiss's (1978) study reportedly spent an average of 41 minutes providing instruction in mathematics. Considering that some time was invariably needed to maintain classroom control, the number of minutes actually spent on instruction was probably less than the reported figure. For those students who found the lessons confusing, lacked the prerequisite skills, or failed to pay attention, the engaged time was probably even less.

Students are less engaged when they are working independently than when they are working in groups (Stevens & Rosenshine, 1981). Weiss reported that teachers spent 43 percent of the time teaching the whole group, 23 percent of the time teaching small groups, and 34 percent of the time supervising students working on individual assignments. It should be stressed that Weiss's data are based on verbal reports by teachers rather than actual observations. Data from observational studies indicate that students may spend as much as two-thirds of the school day working independently (Stallings, Gory, Fairweather, & Needles, 1977; Stallings & Kaskowitz, 1974).

Individualization of instruction is the sine qua non of teaching the learning disabled. Data from Weiss' study (1978) and from Suydam and Osborne's (1977) review suggest that teachers do not individualize instruction for students with learning problems. Commonly cited impediments to individualizing instruction include lack of curriculum materials (Cawley, 1975; Denny, 1978), too little planning time, and the perceived need for excessive record keeping (NACOME, 1975). Unless practical methods are found to assist teachers in individualizing instruction, we can expect little variation from the "tell and drill" method.

MATHEMATICS CURRICULA IN SPECIAL EDUCATION

Mathematics curriculum development in special education has lagged far behind development in regular education (Cawley, 1977). Prompted by the belief that there were no adequate curriculum alternatives for handicapped students, Cawley and his colleagues developed *Project MATH* (Cawley, Fitzmaurice, Goodstein, Lepore, Sedlak, & Althaus, 1976). Another specialized curriculum is *DISTAR Arithmetic* (Englemann & Carnine, 1972, 1975, 1976), which was developed for culturally disadvantaged youngsters and is frequently used in teaching mildly handicapped students. Each of these approaches presents a different view of "what" mathematics should be learned and "how" it should be taught.

Project MATH

Project MATH (Cawley et al., 1976) was designed to provide a comprehensive program in mathematics for mildly handicapped students. The developers began by determining "what" mathematics should be included in the program. Based on the belief that the curriculum must serve behavioral purposes, the program focuses on sets and patterns in levels I and II and on numbers and operations, fractions, geometry, and measurement in levels III and IV (Cawley, 1975). Levels III and IV also include lab activities on such topics as metric measurements and the use of calculators.

Turning their attention to "how" to teach the content, Cawley and his colleagues considered the strengths and weaknesses handicapped children may present and developed an instructional method to compensate for deficient skills. Their goal "was to develop a means that would allow the nonreader to demonstrate proficiency in mathematics without being impeded by a lack of reading ability" (Cawley, 1977, p. 39). To accomplish this goal, they developed what is known as the Interactive Unit.

The Interactive Unit provides the teacher with four methods of instruction: manipulate, display, say, and write. In teaching an objective, the teacher may use one or all of the 16 combinations in the Interactive Unit to meet the needs of individual students.

Project MATH is both a developmental and a remedial program. Placement into the Multiple-Option Curriculum, the developmental portion of the program, is based on performance on the Mathematics Concept Inventory (MCI). The MCI assesses pupils' concepts and skills in the areas of numbers, fractions, geometry, and measurement. Items on the MCI are derived from those that appear in the curriculum, therefore ensuring a close match between assessment and instruction. Students whose performance on the MCI indicates that they have basic concepts and skills in the four operations, fractions, decimals, and/or percentages are placed immediately into the Multiple-Option Curriculum. Students whose performance reveals difficulties in these areas are given the Clinical Mathematics Interview (Cawley, 1978).

The Clinical Mathematics Interview is "an intensive diagnostic procedure that integrates content, mode, and algorithm (or rule)" (Cawley, 1978, p. 224). Students are first given tasks in the write mode. Then they are asked to verbally state the procedure used to solve both correct and incorrect items. Finally, the student completes problems in the other modality combinations included in the Interactive Unit.

Following the Clinical Mathematics Interview, students begin the remedial part of the program. Remedial Modules focus on correcting difficulties with algorithms. Once a student has mastered the necessary remedial modules, the MCI is readministered to determine an appropriate placement into the Multiple-Option Curriculum.

Project MATH can be used with mildly handicapped students from six to sixteen years of age. The maximum grade equivalent of the program is sixth grade (Cawley, 1975).

Project MATH has corrected many of the problems inherent in basal mathematics texts. First, assessment and instruction are linked. Second, progress through the curriculum is based on mastery. Third, specific directions are provided for teaching each objective and criteria for mastery are specified. Fourth, alternatives for individualizing instruction are provided via the Interactive Unit. Fifth, the program has both a developmental and a remedial component.

In commenting on *Project MATH,* Reid and Hresko (1981a) noted that it is "one of the few programs that appears to be useful with adolescent learning disabled students" (p. 308). Since its inception, *Project MATH* has gone through extensive field testing, the results of which have supported its use (Cawley, Fitzmaurice, Shaw, Kahn, & Bates, 1978).

DISTAR Arithmetic

DISTAR Arithmetic (Englemann & Carnine, 1972, 1975, 1976) is comprised of three levels. *DISTAR I* focuses on answering questions, 35 addition facts, backward counting for subtraction problems, word problems, more and less than, and ordinal counting. *DISTAR II* expands on addition and subtraction, introduces multiplication, and includes work on money, time, metric and standard measurement, fractions, and word problems. *DISTAR III* focuses on extending computation in the basic operations and includes word problems as well as formal problem-solving activities.

DISTAR Arithmetic is based on the direct instruction model. In this approach, complex tasks are broken down into component skills, those component skills are taught using behavioral principles, and students are shown how to combine component skills into more complex behaviors (Silbert, Carnine, & Stein, 1981). The program is based on mastery of skills and includes provisions for continuous evaluation of pupil performance.

DISTAR Arithmetic is a carefully designed program that specifies teaching techniques as well as procedures for organizing instruction. Following a placement test on items taken from the curriculum, students are assigned to an appropriate level. The approach is characterized by a teacher working with small groups of students. A teacher presentation book specifies the tasks to be taught and the manner in which the teacher should present them. One commonly used technique is the "Model-Lead-Test-Delayed Test" procedure. Using this technique, a teacher models the correct response, for example, counting by two from 2 to 20. Next, the teacher leads the students through the task. The students are then directed as a group to perform the task. Finally, the teacher calls on individual students to perform the task.

DISTAR Arithmetic makes use of unison responding, immediate feedback, error correction procedures, and frequent praise. With this approach, high levels of student attention are almost ensured, the teacher is provided with continuous information on student progress, and students are reinforced for the tasks they perform.

The effectiveness of *DISTAR Arithmetic* was studied through Project Follow Through (Abt Associates, 1976). The results of that evaluation indicated that students with low socioeconomic backgrounds who had participated in the direct instruction approach performed on the average at the same level as their middle-class peers by the end of the third grade.

While *DISTAR Arithmetic* incorporates much of what is known to be effective instruction, e.g., encouraging high rates of responding, providing feedback and praise, sequencing tasks from least to most complex, it is not without its critics. In commenting on *DISTAR Arithmetic,* Suydam and Osborne (1977) noted that "it is so extreme in controlling the environment and activities for learning that many teachers and mathematics educators find that it conflicts with their beliefs about teaching and the nature of mathematics" (p. 52). Agreeing with this position, Reid and Hresko (1981a) questioned the focus on computational skills. These comments reflect the conflicting opinions of mathematics educators and behaviorists concerning the "what" and the "how" of providing mathematics instruction to students with learning problems.

SPECIALIZED INSTRUCTIONAL APPROACHES

Regular educators have spent a considerable amount of time determining "what" mathematics to teach normally achieving students. Special educators, on the other hand, have focused their attention on determining "how" to provide instruction to students with learning disabilities. As we have seen, one's view of appropriate instruction is in large part determined by one's belief concerning the nature of the learner. In the field of learning disabilities there is no consensus concerning the source of students' learning difficulties. Hence, a variety of instructional approaches have been suggested, each reflecting a particular view concerning the nature of a student's learning problem. These instructional models have been categorized into two types, the basic-ability model and the direct-skills model (Lilly, 1979).

Basic-Ability Model

Advocates of the basic-ability approach subscribe to the belief that learning-disabled students have deficits in one or more underlying abilities that are necessary for learning. Depending upon the theorist, learning problems are viewed as the results of deficits in perceptual-motor functioning (Kephart, 1960), visual-perceptual skills (Frostig, Lefever, & Whittlesey, 1964), or psycholinguistic functioning (Kirk, McCarthy, & Kirk, 1968). Remediation takes one of two forms: (1) weak underlying abilities are directly remediated, or (2) instruction on academic tasks is geared to a student's pattern of underlying strengths and weaknesses (Arter & Jenkins, 1979).

An example of the first method would be to provide visual memory activities for a student who reverses the numeral 3. An example of the second method would be to determine whether a student learned best when information was presented auditorially or visually so that instruction could be geared to the stronger modality.

When this approach is applied to mathematics, students' difficulties are described in terms of deficits in underlying abilities. For example, a student who fails to complete a problem may be viewed as having a figure-ground disturbance. A child who is unable to read a multidigit numeral may be described as having closure difficulties (Bley & Thornton, 1981). Remediation often takes the form of tactile activities such as writing numbers in damp sand (Bley & Thornton, 1981) or using concrete objects.

The basic-ability model is based on the following assumptions: (1) educationally important abilities exist and can be measured, (2) tests used to differentially diagnose learning problems are valid and reliable, (3) weak abilities can be strengthened, (4) remediation of weak abilities enhances academic performance, and (5) academic performance can be improved by matching instruction to a student's strong modality (Arter & Jenkins, 1979).

The validity of these assumptions has been questioned by several educators (Hammill & Larsen, 1974; Hammill & Wiederholt, 1973; Mann, 1971). Contrary to the model, it does not appear that tests used to differentially diagnose learning problems, such as *The Marianne Frostig Developmental Test of Visual Perception* (Frostig et al., 1964) or the *Illinois Test of Psycholinguistic Abilities* (Kirk et al., 1968), correlate very highly with academic performance (Arter & Jenkins, 1979). Nor are these tests judged as possessing sufficient validity or reliability to differentially diagnose learning problems (Salvia & Ysseldyke, 1981). Efforts to train underlying abilities do not reliably result in increasing those abilities, let alone improving academic performance (Hammill & Larsen, 1974; Hammill & Wiederholt, 1973). Moreover, the results of studies that have attempted to match instruction to a student's strong modality have been consistently negative (Arter & Jenkins, 1977).

Despite the lack of research supporting the basic-abilities model, many special educators still cling tenaciously to its tenets. "In a statewide survey in Illinois, it was found that 82% of special education teachers believed that they could, and should, train weak abilities, 99% thought that a child's modality strengths and weaknesses should be a major consideration when devising educational prescriptions, and 93% believed that their students had learned more when they had modified instruction to match modality strengths'' (Arter & Jenkins, 1979, pp. 549–550).

Direct-Skills Model

Proponents of a direct-skills approach are not interested in presumed ''causes'' of learning problems. Instead, they ''adopt the attitude that 'what you see is what it is,' and attempt the shortest and most direct route to the solution of the problem'' (Blankenship & Lilly, 1981, p. 40). The direct-skills model has its basis in behavioral psychology and is known as applied behavior analysis (Haring, Lovitt, Eaton, & Hansen, 1978), responsive teaching (Hall, 1971), or data-based instruction (Lilly, 1979). All of these models may be viewed as variations on a theme; for that reason, the data-based instructional (DBI) model will be used to illustrate the direct-skills approach. This model has been characterized by Blankenship & Lilly (1981) as follows:

1. DBI focuses on the specific behavior of concern, and essentially ignores the underlying causes of the behavior. This focus on observable behavior makes coordination between special and regular education an easier task, since both sides are talking the same language.
2. DBI avoids categorical labels, since the focus is on definition of academic and social problems in behavioral terms, thus leading to direct measurement and instructional intervention without necessity of further labeling.
3. DBI emphasizes functional assessment within the instructional setting as contrasted with norm-referenced measurement which only incidentally relates to the problems being encountered by an individual student and his/her teacher.
4. DBI emphasizes the use of specific instructional objectives as the basis for teaching, which means that instruction is well focused and expected outcomes can be agreed upon by parents, teachers, and other school officials.
5. DBI does not rely on a specific teaching methodology, but rather provides for evaluation of instructional effectiveness in each individual case. A teaching approach is not good or bad in a general sense, but might work with one student and not with another. These decisions are made not on the basis of subjective teacher judgment, but rather from data on student progress. Arguments over which teaching technique is best are rendered obsolete.
6. DBI is an individualized instruction model which operates on the basis of individualized instructional objectives and progress monitor-

ing, but not necessarily one-on-one teaching. Instruction can be in groups, if it is effective in moving students toward stated instructional objectives.

7. DBI stresses collection of continuous data on student progress, so that teachers need not wait six weeks, two months, or an academic year to determine whether an instructional intervention is a success or failure.
8. If a student does not achieve an *instructional objective,* DBI focuses the teacher's attention on inadequacies in the instructional program, not inadequacies in the student.
9. DBI fits perfectly with the IEP requirements of Public Law 94-142, in that any teacher who has implemented DBI will have readily available all required instructional elements of the IEP. (pp. 40–41)

The DBI approach does not specify "what" mathematics to teach. When using this approach to teach mathematics to students classified as learning disabled, assessment is based on the mathematics textbook used in a student's regular classroom. Depending upon a student's performance, a decision is made to (1) keep the student in the regular curriculum material at a level commensurate with his or her skills, (2) select other commercial or teacher-made materials to teach necessary concepts and skills that upon mastery will allow the student to reenter the regular curriculum, or (3) place the student in an alternate curriculum that emphasizes functional mathematical skills.

To illustrate the DBI approach to mathematics, a sample project is provided followed by a discussion of research related to the use of DBI techniques to improve the mathematical performance of learning-disabled students.

Improving Performance in Solving Word Problems*

Number of Students

Seven students participated in the project, all of whom were classified as learning disabled by the school district, which referred them to the Curriculum Research Classroom, Experimental Education Unit, University of Washington. For the purposes of this example, data will be provided for one student whose performance was typical of the group.

---

*Substantial portions of this section are excerpted, with minor modifications from the following text: Blankenship, C., & Lilly, M.S. *Mainstreaming students with learning and behavior problems: Techniques for the classroom teacher*. New York: Holt, Rinehart & Winston, 1981, pp. 279–284.

Setting

Curriculum Research Classroom.

Target Behavior/Aim

Increase accuracy in computing arithmetic word problems of the type $n - 9 = 3$ and $12 - n = 3$ to 100 percent correct for three days.

Definition of Target Behavior

Problems not attempted and wrong answers were counted as incorrect. Correct answers corresponded to the answer one would obtain by performing the computation presented in the word problem.

Materials

For each of four classes of word problems taught, the following materials were developed:

1. Demonstration Sheets—Five demonstration sheets, each containing two sample problems, one of each type to be instructed. A sample demonstration sheet for Class 4 word problems ($n - 2 = 2$ and $7 - n = 4$) is shown in Exhibit 2–1.
2. Teacher Cue Sheets—Two scripts were written for each problem type to be instructed. A sample cue sheet for one type of Class 4 problem ($n - 2 = 2$) is shown in Exhibit 2–2.
3. Correction Sheets—Worksheets were developed which contained blank equation blocks and which were labeled to correspond to each type of problem taught.
4. Daily Worksheets—Five worksheets per class of problems were developed, each of which contained ten problems of each of two problem types to be instructed. The same problems were presented each day, but the number facts varied from one set to another. The problems were presented in a different order each day.

Observation System and Measurement Technique

Permanent product recording was used to measure the number of problems answered correctly and incorrectly per day. The student's performance was also timed, thus enabling the teacher to record rate data. Both percent of problems computed per day and problems per minute were charted.

Exhibit 2–1 Demonstration Sheet Class 4 Form a

Sam has 2 frogs. He has 2 more ducks than frogs. How many ducks does Sam have? ____________

Size of Larger Group	Action	Size of Smaller Group		Difference
☐	☐	☐	=	☐

Mrs. Hill has 7 plates. She has 4 more plates than glasses. How many glasses does she have? ____________

Size of Larger Group	Action	Size of Smaller Group		Difference
☐	☐	☐	=	☐

Procedures

During baseline in Class 4 problems, Toni was asked to compute twenty word problems as best he could. No instruction or feedback was provided. The first intervention consisted of the teacher demonstrating how to solve problems of the type $n - 9 = 3$ and $12 - n = 3$ using the appropriate Demonstration and Teacher Cue Sheets. No feedback was provided to the student concerning his accuracy in computing the problems.

During the second intervention the demonstration was discontinued and the student was required to correct erred problems with the teacher. The student read each incorrectly answered problem and the teacher led him through the process by asking the questions contained on the Teacher Cue Sheet.

The third intervention consisted of a combination of Techniques 1 and 2. The demonstration was provided as in Technique 1 and the student corrected erred problems as described in Technique 2. During the fourth intervention, the student was given the opportunity to earn a notebook if he scored 100 percent correct for three days.

Exhibit 2–2 Cue Sheet Class 4

Student reads orally the sample comparison addition problem shown in Exhibit 2–1. Teacher asks the following questions, supplies answers only if student cannot, and directs the student to place numbers in appropriate spaces in the equation blocks on demonstration sheet.

1. What does the problem ask you to find? (size of larger group)
2. Do you know the size of the larger group? (no)
3. So we leave this box blank.
4. Do you know the size of the smaller group? (yes)
5. What is the size of the smaller group? (2 frogs)
6. We put the size of the smaller group right here.
7. Do you know the difference between the size of the larger group and the size of the smaller group? (yes)
8. What is the difference between the size of the larger group and the size of the smaller group? (2)
9. We put the difference between the size of the larger group and the size of the smaller group right here.
10. What is the action? Joining, separating, or comparing? (comparing)
11. What sign do we use for comparing? (−)
12. We put the minus sign right here.
13. Now let's solve the equation. (Student reads, teacher corrects verbally.)
14. To find the size of the larger group we add the size of the smaller group to the difference.
15. Please add the size of the smaller group to the difference.
16. What is the size of the larger group? (4 ducks)
17. We put the size of the larger group right here.
18. Student reads number sequence.
19. Is that number sentence true? (If not, student corrects it.)
20. Teacher reads problem question, student answers with noun and number.

Results

Toni's daily data are shown in Figure 2–2. During baseline Toni's accuracy ranged from 0 percent to 25 percent correct. His average accuracy during baseline and the first intervention was 14.2 percent and 11.7 percent correct, respectively. During the second intervention his accuracy ranged from 0 percent to 70 percent correct with an average of about 32 percent. After the first few days of Technique 3, his scores varied between 85 percent and 100 percent correct. By the sixth day of the fourth intervention, Toni reached the desired aim.

Comment

While a combination of demonstration and correction was effective in increasing Toni's accuracy, it did not result in his achieving the desired aim. Toni's performance during baseline and Technique 1 provides an example of what we mean by a student who "can't" or at least doesn't perform a skill very accurately.

Figure 2–2 Toni's Percent Correct Scores on Word Problems

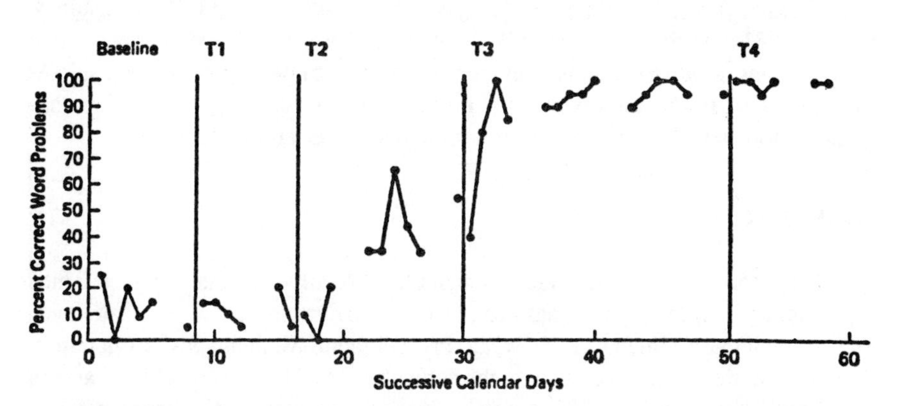

Source: Blankenship, C., and Lilly, M.S. (1981). *Mainstreaming students with learning and behavior problems: Techniques for the classroom teacher.* New York: Holt, Rinehart & Winston. Used by permission.

While Technique 2 originally looked promising, performance eventually leveled off. During Technique 3, Toni's performance is typical of a student who "can" but doesn't perform very consistently. In this case the chance to earn a notebook provided an appropriate incentive for Toni to master this class of word problems.

Although these techniques were used with individual students, the demonstration and correction procedures could easily be adapted to small groups of students.

Techniques for Remediation in Mathematics

A variety of techniques have been used to remediate common difficulties of learning-disabled students in mathematics. Some of the techniques employed include (1) instructions, (2) demonstration, (3) modeling, (4) drill and practice, (5) prompts, (6) fading and shaping, (7) feedback, (8) reinforcement, and (9) contingencies for errors (Haring et al., 1978). Improvements in the computational performance of learning-disabled students have resulted from requiring them to verbalize problems and their answers before writing them (Lovitt & Curtiss, 1968). Similarly, demonstration and modeling plus feedback has been used to reduce systematic errors in subtraction (Blankenship, 1978). Computational speed has also been increased through the use of prompts to encourage students to work faster (Smith & Lovitt, 1975). Various types of reinforcement techniques have been used to increase computational accuracy (Lovitt, 1978). Recently, attention has been focused on increasing learning-disabled students' abilities to generalize computational skills through the use of verbal reminders (Baumgartner, 1979) and reinforcement (Blankenship & Baumgartner, 1982).

The majority of research using a direct-skills approach has focused on computational skills. The effectiveness of the model is well-documented (Blankenship & Lilly, 1981; Haring et al., 1978). However, there have been relatively few applications to other mathematical areas such as word problems (Blankenship & Lovitt, 1974, 1976). Future research must necessarily address a variety of mathematical topics including estimation, problem solving, and measurement.

SUMMARY

Many learning-disabled students encounter difficulties in mathematics. Until quite recently little emphasis has been placed on determining "what" mathematics to teach these students or "how" to instruct them. Mathematics instruction in regular classrooms is dependent on the content of basal texts from which teachers pick and choose topics. Unfortunately, because of this practice computational skills are stressed while other mathematical topics are ignored.

Standard curriculum materials pose a variety of problems for learning-disabled students. The readability levels of some texts prevent them from independently following directions or computing application problems. Often too few examples are provided to ensure mastery. To circumvent these problems, teachers must individualize instruction. Yet few teachers actually do individualize instruction in mathematics for students with special needs.

To improve mathematics curriculum and instruction for the learning disabled, a national effort is required (Cawley, 1981). In the coming years, attention must be focused on

1. Assessing the mathematical skills of learning-disabled students. No large-scale assessment has ever been conducted with the learning disabled.
2. Training regular educators to individualize instruction for students with special needs.
3. The development and refinement of alternative curricula for students who are unable to make progress in the standard curriculum.
4. Research designed to identify effective instructional techniques to improve the mathematical skills of learning-disabled students.

REFERENCES

Abt Associates. (1976). Education as experimentation: A planned variation model (Vol. 3A). Cambridge, Mass.: Author.

Arter, J.A., & Jenkins, J.R. (1977). Examining the benefits and prevalence of modality considerations in special education. *Journal of Special Education, 11*, 281–298.

Arter, J.A., & Jenkins, J.R. (1979). Differential diagnosis-prescriptive teaching: A critical appraisal. *Review of Educational Research, 49*, 517–555.

Baumgartner, M. (1979). *Generalization of improved subtraction regrouping skills from resource rooms to the regular class.* Unpublished doctoral dissertation, University of Illinois.

Blankenship, C.S. (1978). Remediating systematic inversion errors in subtraction through the use of demonstration and feedback. *Learning Disability Quarterly, 1,* 12–22.

Blankenship, C.S., & Baumgartner, M. (1982). Programming generalization of computational skills. *Learning Disability Quarterly, 5,* 152–162.

Blankenship, C.S., & Lilly, M.S. (1981). *Mainstreaming students with learning problems: Techniques for the classroom teacher.* New York: Holt, Rinehart & Winston.

Blankenship, C.S., & Lovitt, T.C. (1974). *Story problem data collected in curriculum research classroom.* Seattle: University of Washington, Experimental Education Unit.

Blankenship, C.S., & Lovitt, T.C. (1976). Story problems: Merely confusing or downright befuddling. *Journal for Research in Mathematics Education, 7,* 290–298.

Bley, N.S., & Thornton, C.A. (1981). *Teaching mathematics to the learning disabled.* Rockville, Md.: Aspen Systems.

Bruner, J.S. (1963). *The process of education.* New York: Vintage Books.

Cawley, J.F. (1975). Special education: Selected issues and innovations. In A.D. Roberts (Ed.), *Educational innovation: Alternatives in curriculum and instruction.* Boston: Allyn & Bacon.

Cawley, J.F. (1977). Curriculum: One perspective for special education. In R.D. Kneedler & S. Tarver (Eds.), *Changing perspectives in special education.* Columbus, Ohio: Charles E. Merrill.

Cawley, J.F. (1978). An instructional design in mathematics. In C. Mann, C. Goodman, & J.C. Wiederholt (Eds.), *Teaching the learning-disabled adolescent.* Boston: Houghton Mifflin.

Cawley, J.F. (1981). Commentary. *Topics in Learning and Learning Disabilities, 1,* 89–95.

Cawley, J.F., Fitzmaurice, A.M., Shaw, R.A., Kahn, H., & Bates, H. (1978). Mathematics and learning disabled youth: The upper grade levels. *Learning Disability Quarterly, 1,* 37–52.

Cawley, J.F., Goodstein, H.A., Fitzmaurice, A.M., Lepore, A.V., Sedlak, R., & Althaus, V. (1976). *Project MATH.* Tulsa, Okla.: Educational Progress Corporation.

Denholm, R.A., Hankins, D.D., Herrick, M.C., & Vojtka, G.R. (1977). *Mathematics for individual achievement.* Atlanta: Houghton Mifflin.

Denny, T. (1978). Some still do: River Acres, Texas. (1978). In R.E. Stake, J.A. Easley, Jr., & collaborators. *Case studies in science education.* University of Illinois at Urbana-Champaign, College of Education. (Stock No. 038-000-0377-1, 038-000-003763, and 038-000-00383-6). Washington, D.C.: U.S. Government Printing Office.

Educational Products Information Exchange. (1976, December 1). Research findings: NSAIM: Two years later. *EPIEgram, 5*(5).

Englemann, S., & Carnine, D. (1972). *DISTAR: Arithmetic level III.* Chicago: Science Research Associates.

Englemann, S., & Carnine, D. (1975). *DISTAR: Arithmetic level I.* Chicago: Science Research Associates.

Englemann, S., & Carnine, D. (1976). *DISTAR: Arithmetic level II.* Chicago: Science Research Associates.

Fey, J.T. (1980). Mathematics teaching today: Perspectives from three national surveys for the elementary grades. In National Science Foundation Report, *What are the needs in precollege science, mathematics, and social science education? Views from the field.* Washington, D.C.: U.S. Government Printing Office.

Frostig, M., Lefever, D., & Whittlesey, J. (1964). *The Marianne Frostig Developmental Test of Visual Perception.* Palo Alto, Calif.: Consulting Psychology Press.

Gagné, R.M., & Briggs, L.J. (1974). *Principles of instructional design*. New York: Holt, Rinehart & Winston.

Glennon, V.J. (1965). . . . And now synthesis: A theoretical model for mathematics education. *The Arithmetic Teacher, 12*, 134–141.

Glennon, V.J. (1975). Elementary school mathematics: Alternatives and imperatives. In A.D. Roberts (Ed.), *Educational innovation: Alternatives in curriculum and instruction*. Boston: Allyn & Bacon.

Hall, R.V. (1971). *Managing behavior, Part 1: The measurement of behavior*. Lawrence, Kans.: H & H Enterprises.

Hammill, D.D., & Larsen, S.C. (1974). The effectiveness of psycholinguistic training. *Exceptional Children, 41*, 5–14.

Hammill, D.D., & Wiederholt, J.L. (1973). Review of the Frostig Visual Perception Test and the related training program. In L. Mann and D.A. Sabatino (Eds.), *The first review of special education* (Vol. 1). Philadelphia: JSE Press.

Haring, N.G., Lovitt, T.C., Eaton, M.D., & Hansen, C.L. (Eds.). (1978). *The fourth R: Research in the classroom*. Columbus, Ohio: Charles E. Merrill.

Houck, C., & Given, B. (1981). Status of SLD programs: Indications from a teacher survey. *Learning Disability Quarterly, 4*, 320–325.

Kephart, N.C. (1960). *The slow learner in the classroom*. Columbus, Ohio: Charles E. Merrill.

Kirk, S.A., McCarthy, L., & Kirk, W.D. (1968). *Illinois Test of Psycholinguistic Abilities* (rev. ed.). Urbana, Ill.: University of Illinois Press.

Lilly, M.S. (Ed.). (1979). *Children with exceptional needs: A survey of special education*. New York: Holt, Rinehart & Winston.

Lovitt, T.C. (1978). Arithmetic. In N.G. Haring, T.C. Lovitt, M.D. Eaton, & C.L. Hansen (Eds.), *The fourth R: Research in the classroom*. Columbus, Ohio: Charles E. Merrill.

Lovitt, T.C., & Curtiss, K.A. (1968). Effects of manipulating an antecedent event on mathematics response rate. *Journal of Applied Behavior Analysis, 1*, 329–333.

Mann, L. (1971). Psychometric phrenology and the new faculty psychology: The case against ability and assessment training. *Journal of Special Education, 5*, 3–65.

McLeod, T.M., & Armstrong, S.W. (1982). Learning disabilities in mathematics—skill deficits and remedial approaches at the intermediate and secondary level. *Learning Disability Quarterly, 5*, 305–311.

National Advisory Committee on Mathematical Education. *Overview and analysis of school mathematics grades K–12*. (1975). Washington, D.C.: Conference Board of the Mathematical Sciences.

Piaget, J. (1973). Comments on mathematical education. In A.G. Howson (Ed.), *Developments in mathematical education*. London: Cambridge University Press.

Reid, D.K., & Hresko, W.P. (1981a). *A cognitive approach to learning disabilities*. New York: McGraw-Hill.

Reid, D.K., & Hresko, W.P. (1981b). From the editors. *Topics in Learning and Learning Disabilities, 1*, viii–ix.

Robitaille, D.F., & Dirks, M.F. (1980). *Curriculum models in mathematics*. Unpublished manuscript, University of British Columbia.

Robitaille, D.F., & Dirks, M.F. (1982). Models for the mathematics curriculum. *For the Learning of Mathematics, 2*(3).

Salvia, J., & Ysseldyke, J.E. (1981). *Assessment in special and remedial education* (2nd ed). Boston: Houghton Mifflin.

Silbert, J., Carnine, D., & Stein, M. (1981). *Direct instruction mathematics*. Columbus, Ohio: Charles E. Merrill.

Skager, R.W. (1969, February). *Student entry skills and the evaluation of instructional programs: A case study*. Paper presented at the meeting of the American Educational Research Association, Los Angeles, Calif. (ERIC Document Reproduction Service No. ED 029 364)

Smith, D.D., & Lovitt, T.C. (1975). The use of modeling techniques to influence the acquisition of computational arithmetic skills in learning disabled children. In E. Ramp and G. Semb (Eds.), *Behavior analysis: Areas of research and application*. Englewood Cliffs, N.J.: Prentice-Hall.

Stake, R.E., & Easley, J.A., Jr., & collaborators. (1978). *Case studies in science education*. University of Illinois at Urbana-Champaign, College of Education. (Stock No. 038-000-0377-1, 038-000-003763, and 038-000-00383-6). Washington, D.C.: U.S. Government Printing Office.

Stallings, J., Gory, R., Fairweather, J., & Needles, M. (1977). *Early childhood education classroom evaluation*. Menlo Park, Calif.: Stanford Research Institute.

Stallings, J., & Kaskowitz, D. (1974). *Follow through classroom observation evaluation, 1972–73*. Menlo Park, Calif.: Stanford Research Institute.

Stevens, R., & Rosenshine, B. (1981). Advances in research on teaching. *Exceptional Education Quarterly, 2,* 1–9.

Suydam, M.N., & Osborne, A. (1977). *The status of pre-college science, mathematics, and social science education: 1955–1975. Vol. II: Mathematics education*. Columbus: The Ohio State University Center for Science and Mathematics Education.

Thorndike, E.L. (1924). *The psychology of arithmetic*. New York: Macmillan.

Trafton, P.R. (1981). Assessing the mathematics curriculum today. In M.M. Lindquist (Ed.), *Selected issues in mathematics education*. Berkeley: McCutchan.

U.S. Department of Education (1980). *Second annual report to Congress on the implementation of Public Law 94-142: The Education for all Handicapped Children Act*. Washington, D.C.: U.S. Government Printing Office.

Weiss, I. (1978). *Report of the 1977 national survey of science, mathematics, and social studies education*. Research Triangle Park, N.C.: Research Triangle Institute.

Welch, W. (1978). Urbanville. In R.E. Stake & J.A. Easley, Jr., & collaborators. *Case studies in science education*. University of Illinois at Urbana-Champaign, College of Education. (Stock No. 038-000-0377-1, 038-000-003763, and 038-000-00383-6). Washington, D.C.: U.S. Government Printing Office.

Chapter 3

Assessment: Examination and Utilization from Pre-K through Secondary Levels

Henry A. Goodstein

Public Law 94-142 provides for the systematic assessment of learners that might require special education programming. The major purposes of this assessment process are (1) to identify, diagnose, or label students according to handicapping condition and (2) to aid in individual program planning (Myers & Hammill, 1982). The focus of this chapter is on assessment practices and instruments designed to determine the nature and degree of mathematics disability among children with specific learning disabilities. The emphasis will be on assessment for the purpose of assisting in instructional planning in contrast to diagnosis.

The resolution of conceptual and practical difficulties in the design of assessment strategies that might differentiate categories of underachievement in mathematics (e.g., Cawley, Fitzmaurice, Shaw, Kahn, & Bates, 1979) is an important issue. However, to date, this unresolved issue does not appear to have had a significant influence on assessment practices (Berk, 1981). A more important influence derives from the conceptual views of mathematics and learning disabilities held by the person responsible for designing the assessment materials.

ALTERNATIVE ASSESSMENT MODELS

McEntire (1981) suggested three prominent views of the relationship between mathematics and learning disabilities. Those who view a relationship between neurological-mental ability deficits and mathematics disability (e.g., Johnson, 1979) would favor the inclusion of traditional psychological process/ability tests (e.g., *Illinois Test of Psycholinguistic Ability* (Kirk, McCarthy, & Kirk, 1968)) in their assessment strategies. Those who favor the view that cognitive-developmental differences are closely associated with mathematics disability (e.g., Kamii, 1981; Reid & Hresko, 1981) would rely more on the administration of Piagetian tasks as well as learner interviews.

Those whose orientation is more behavioral and content/task-determined would tend to reject assessment instruments or techniques that sought to describe the learner's status except in reference to the mathematics task and its direct prerequisites. While it is clear that those who advocate the model of applied behavior analysis (e.g., Lovitt, 1975) have little use for consideration of developmental characteristics of the learner, the boundary between the first two views is not so distinct.

This is especially true for the mathematics educator who is interested in preventing and remediating mathematics disabilities. Reisman's assessment model (Reisman, 1982) is influenced by the work of Bruner (1960), Piaget (1932), and Gagné (1977). Underhill, Uprichard, and Heddens (1980), while relying heavily upon the research into learning hierarchies by Gagné (1977), have integrated a developmental orientation in their task-process model of assessment. Glennon and Wilson's (1972) diagnostic-prescriptive teaching model is heavily influenced by Bruner in its design and application. What does, perhaps, distinguish these mathematics educators is the centrality of the mathematics, its structure and cognitive demands, in consideration of the process of assessment.

DEFINITION OF ASSESSMENT

Assessment may be defined as the systematic process of determining the current status of a learner's knowledge and skill and identifying any additional information requisite for effective instructional planning. Assessment is not synonymous with measurement. It includes both quantitative descriptions of the learner (measurement) and qualitative descriptions (nonmeasurement). In addition, the assessment process will always include value judgments, as both quantitative and qualitative descriptions are weighted and synthesized (Gronlund, 1981). While the available measurement devices to aid assessment in mathematics continue to increase and improve in quality, tests alone will never prove adequate for the assessment task.

Definition of a Test

Administration of tests is the most common means of obtaining quantitative descriptions of learner performance in mathematics. A test is defined as any standardized sample of stimuli (items) intended to elicit responses under standardized conditions. This definition does not restrict tests to paper-and-pencil instruments. Neither does it necessarily distinguish between commercially distributed and teacher-constructed tests. The term *informal test* is often used in descriptions of teacher-constructed tests. By inference, informal often suggests a lack of either standardization of items or conditions of administration, minimum requirements for reliable quantitative description of a learner. Informal tests lacking such standardization would best be considered qualitative descriptions.

Norm-Referenced Measurement

Test scores are interpreted with reference to either of two measurement scales, relative or absolute. Tests that yield scores interpreted on a relative measurement scale are termed *norm-referenced.* Raw scores are converted or transformed to derived scores that allow ease of comparison of the learner's performance with that of some clearly defined reference group, i.e., the norm. Examples of such transformed scores include grade or age equivalents, standard scores, percentiles, and stanines.

For norm-referenced tests the total raw score typically has little meaning in any absolute sense. The score obtains meaning by comparison with scores obtained by other students. While the item pool from which test items will ultimately be selected is presumably developed with respect to a content-valid set of specifications, the actual test items will be selected on the basis of certain statistical, psychometric properties. For most norm-referenced tests, items will be favored that are of mid-range difficulty and contribute to the discrimination of good and poor performers on the test as a whole. Tests constructed in this manner provide for maximum differentiation among students. Such differentiation contributes to the reliable interpretation of norm-referenced test scores.

Criterion-Referenced Measurement

Tests that yield scores interpreted on an absolute measurement scale are termed domain-referenced or more commonly *criterion-referenced.* Raw scores are typically converted to percentages. Meaning is explicitly or implicitly conferred to the score with respect to some domain of performance. That is, the score is viewed as an estimate of performance that the student would have obtained if he or she were administered all similar items from the domain. Reliable and valid test scores are directly related to the clarity of domain specifications and representative selection of items from the domain (Goodstein, 1982).

A cutoff or advancement score (often referred to as a criterion score) is a feature associated with many criterion-referenced tests. This is a score suggested by the test author that is the minimum level of mastery of the domain represented by the test items. Most of the cutoff scores suggested are arbitrary, as opposed to resulting from empirical validation. It should be stressed that a well-constructed criterion-referenced test may not feature a suggested cutoff score. Conversely, a collection of test items (in the absence of clear domain specifications) with an associated "criterion-score" does not constitute a criterion-referenced test.

Most criterion-referenced tests will describe the item domain with a behavioral or instructional objective. However, a chart or table linking test items with instructional objectives (while potentially useful in suggesting possible additional assessment tactics) does not necessarily indicate that the test is criterion-referenced. Criterion-referenced tests must provide a sufficient number of items to reliably assess each objective (domain). Test items will be selected as representative of the range of items in the domain (generally without regard to their difficulty or discrimination indices). Norm-referenced tests will tend to sample more objectives with fewer items. Criterion-referenced tests will tend to sample fewer objectives (on a single test form) with more items.

In recent years, some test publishers have attempted to capitalize on the interest in criterion-referenced testing by linking test items on their norm-referenced test with instructional objectives. However, since this often results in as few as a single item to assess an objective, reliable interpretation is problematic at best. It should be noted that the opposite tactic is possible (and in evidence)—the development of norms for criterion-referenced tests.

Both norm-referenced and criterion-referenced measurements serve legitimate purposes in the assessment of mathematics disabilities. Few would advocate exclusive reliance upon norm-referenced measurements. However, the number of advocates of the exclusive use of criterion-referenced measurement systems appears destined to grow. As criterion-referenced testing systems become more comprehensive and expertly crafted, the initial step in many assessment strategies of administering a norm-referenced instrument solely to document the degree of disability may be viewed as lacking utility.

READINESS TESTING

Readiness testing is directed toward preventing serious mathematics disabilities *before* they occur. It is in testing readiness that divergent views of the centrality of various factors associated with disability in mathematics significantly influence the assessment program. Those stressing cognitive-developmental differences might rely heavily upon the Piagetian interview (Kagan, 1976) or the administration of a series of Piagetian tasks, as suggested by Reisman (1982). Those viewing disabilities as neurological-mental deficits would select various psychoeducational tests and traditional readiness tests. Scores would be interpreted as indicating sufficient or insufficient aptitude for formal instruction in mathematics.

Those with an educational task orientation would either select a criterion-referenced inventory of skills or concepts that presumably are prerequisite to beginning mathematics instruction or a norm-referenced readiness test that surveyed a comparable range of tasks. These prerequisite skills or concepts would logically relate to the structure of the mathematics to be learned, as opposed to a general psychological construct (e.g., spatial perception) thought to be related to mathematics learning.

Readiness testing ought to be an integral component of assessment of mathematics disabilities. To the extent that learners with potential mathematics disabilities may be identified, early intervention can be designed to minimize potential disability. A substantial number of norm-referenced readiness tests are available from test publishers. Consistent with their intended use as screening instruments, most readiness tests are designed for group administration. The minimization of reading skill requirements by the use of pictorial items, oral directions, and simple formats for recording responses tend to offset the typical disadvantages of group administration for children with learning disabilities. For learners who are unresponsive in group settings, a readiness test designed for group testing could be administered on an individual basis (provided that efforts were made to follow the standardized directions).

Norm-Referenced Tests

The following are brief descriptions of several popular norm-referenced readiness tests that may be useful in identifying potential mathematics disabilities.

• The *Boehm Test of Basic Concepts* (BTBC) (Boehm, 1971) is intended for use with children in kindergarten, grade 1, and grade 2. It is designed to measure understanding of 50 basic concepts considered necessary in the first years of school. Each item consists of a set of pictures. The learner is asked to mark the picture that illustrates the selected concept. Concepts are classified into four context categories. Eighteen of the 50 concepts are classified in the quantity category. The author admits that the classification of BTBC concepts is somewhat arbitrary and alternative classifications could be suggested. Unfortunately, no normative data is available for context category scores.

• The revised *Metropolitan Readiness Tests* (MRT) (Nurse & McGauvran, 1976) provides a survey of some important skills thought to be needed for success in beginning mathematics. The MRT is divided into two levels. Level I is intended to function as a reading readiness test designed for use from the beginning through the middle of kindergarten. Quantitative Language is a subtest included in the skill area of Language. It measures an understanding of certain basic quantitative concepts such as size, shape, and number-quantity relationships.

Level II is designed for use from the middle of kindergarten to the beginning of first grade. It includes a quantitative skill area as an optional assessment. The skill area includes subtests of quantitative concepts and quantitative operations. Quantitative concepts include number-numeral correspondence, conservation, part-whole relationships, and quantitative reasoning. Quantitative operations include counting and addition and subtraction of sets.

• The *Stanford Early School Achievement Test* (SESAT) (Madden, Gardiner, & Collins, 1981) is an integral component of the *Stanford Achievement Test Series* (Gardiner, Rudman, Karlesen, & Merewin, 1981). It is divided into two levels: SESAT 1 for grades K.0–K.9 and SESAT 2 for grades K.5–1.9. At Level 1, the mathematics subtest is divided into two parts. The second part, which is optional, is recommended for use only with children who have been introduced to concrete experiences in mathematics. SESAT measures basic number concepts, including number and numeration, counting, and the language of comparison; knowledge of geometric shapes; principles of measurement; understanding the language of simple mathematical problem situations; and basic addition and subtraction facts (sums to 9).

• The *Beginning Educational Assessment* (BEA) (Cawley, Cawley, Cherkes, & Fitzmaurice, 1980) is an integral component of the *Scott-Foresman Achievement Series* (Wick & Smith, 1980), comprising Levels 4, 5, and 6. The purpose of the BEA is to evaluate the development of learning processes and capabilities of children ages four through six. The subtests vary in composition at each level to reflect the continually growing repertory of learning processes as the child develops. One subtest is common for all three levels—Developmental Mathematics. The subtest uses an unusual format of two pairs of stimuli (numerals, geometric shapes, sets, etc.) with the child asked to select the pairs that are most alike. A mathematics subtest included at Level 6 is an achievement test measuring both computation skills and performance on orally presented problems.

• The *Tests of Basic Experiences 2* (TOBE2) (Moss, 1978) is designed to assess important concepts that contribute to a child's preparedness for school. TOBE2 is organized into two levels. Level K is intended for preschool and kindergarten and Level L for kindergarten and first grade. The mathematics subtest assesses quantitative vocabulary and concepts, geometric shapes, and money.

Criterion-Referenced Tests

• The *Brigance Diagnostic Inventory of Early Development* (DIED) (Brigance, 1978) is a criterion-referenced inventory intended for use with individuals with developmental ages less than seven years. The DIED includes 12 math sequences (criterion-referenced tests) to assess mathematics readiness skills such as rote counting, writing dictated numerals, and recognition of money. In contrast to the norm-referenced tests, which are designed for group administration, the DIED is designed for individual administration.

Aptitude vs. Achievement

Tests can be classified as aptitude or achievement tests. Jensen (1980) has suggested that the classification be determined on the basis of sampling, time, function, and stability. Aptitude tests in contrast with achievement tests (1) sample a broader and more heterogeneous set of experiences, (2) cover accumulated knowledge and skills from all past experience, (3) are designed for prediction of future achievement, and (4) are less susceptible to the effects of instruction.

The classification of readiness tests with respect to mathematics content is at best problematic. The content of many readiness tests appears to be a blending of concepts that could be acquired by the child prior to formal instruction in a mathematics program and skills or concepts that are an integral component of formal programs. As formal mathematics programs continue to extend their content into "readiness" activities and more learners participate in early education programs, readiness tests will more and more function as early achievement tests.

GROUP ACHIEVEMENT TESTS: ELEMENTARY

Assessment of mathematics disability in the elementary school typically initiates with the teacher identifying a learner who is failing to make adequate progress in the formal mathematics curriculum. A data source often sought to confirm poor performance is the learner's scores on the group achievement test battery administered annually or biannually by the school system. The mathematics subtests of these batteries are intended as broad survey tests. As such, no attempt is made to measure all significant concepts and skills. However, the test items are typically of high quality and the resulting test scores are reliable and valid norm-referenced measurements.

Occasionally, a group-administered achievement test might be used in classification or placement decisions. While generally inappropriate, this could be justified if the test was individually administered and the group test selected was judged to include a more complete behavior sample than an individual achievement test (Salvia & Ysseldyke, 1981).

Since test items were selected at the mid-range of difficulty for the normative population, learners with mathematics disabilities would be expected to miss many items. To the extent that information as to what items the learner could answer correctly would be as valuable as information about items failed, survey achievement test scores are only partially useful.

Norm-Referenced Tests

The following are brief descriptions of the mathematics subtests of some of the popular norm-referenced achievement tests used in the elementary grades.

- The *Stanford Achievement Tests* (SAT) (Gardiner, Rudman, Karlesen, & Merewin, 1981) is divided into three primary levels intended for use from the middle of first grade through the end of fourth grade and two intermediate levels intended for use from the middle of fourth grade through the end of seventh grade. At the first primary level the mathematics section is divided between concepts of number and mathematics computation and applications subtests. At successive levels, the mathematics computation and mathematics applications subtests are separate.

The content of the subtests at each level reflects typical scope and sequence mathematics objectives. For example, at Primary 3 (grade level 3.5–4.9), concepts of number include naming numbers, reading numbers, place value, and ordering, as well as concepts of fractional numbers. Mathematics computation includes addition and subtraction of whole numbers, with and without renaming, and multiplication and division examples. Mathematics applications includes story problems, reading and interpreting graphs, and principles of geometry and measurement.

• The *California Achievement Tests* (CAT) (CTB/McGraw-Hill, 1977) is divided into seven levels that span from the beginning of first grade (Level 11) to the end of seventh grade (Level 17). A readiness test, Level 10, and two tests, Levels 18 and 19, intended for grades 7–12 are also part of the series. The mathematics section includes a mathematics computation and a mathematics concepts and applications subtest. Mathematics computation assesses skill development in the solution of addition, subtraction, multiplication, and division problems involving whole numbers, fractions, mixed numbers, decimals, and algebraic expressions. The mathematics concepts and applications subtests include items that require students to recognize concepts and apply problem-solving operations in various contexts including numeration, number theory, number sentences, number properties, common scales, geometry, measurement, graphs, and story problems.

• The *Metropolitan Achievement Tests* (MAT) (Prescott, Balow, Hogan, & Farr, 1978) is a two-component system of achievement evaluation. The survey component is similar to traditional group achievement batteries. The instructional component assesses selected objectives within subject matter domains (e.g., mathematics), with at least three items for each objective. While norm-referenced, the instructional component tests are designed consistent with their intended use as a criterion-referenced test.

The MAT Survey Test Battery is divided into nine levels, the first seven of which span grade levels from kindergarten (Preprimer) to the end of sixth grade (Intermediate). The new MAT has a single mathematics subtest with items assessing objectives within seven "strands:" numeration, geometry and measurement, problem solving, operations-whole numbers, operations-laws and properties, operations-fractions and decimals, and graphs and statistics. The survey tests assess each objective with a single item.

• The *Iowa Tests of Basic Skills* (ITBS) (Hieronymous, Lindquist, & Hoover, 1979) is a group achievement test battery intended for use in grades kindergarten through grade 9. The battery is divided into six levels. Two levels constitute the Early Primary Battery, ranging from grade K.1 to 1.9; two levels, the Primary Battery, from grade 1.7 to 3.5; and, six levels, the Multilevel Battery, from grade 3 to 8.9.

Mathematics assessment is composed of three subtests. Mathematics Concepts surveys objectives from the following areas: numeration, number systems, and sets; equations, inequalities, and number systems; whole numbers and integers; fractions; decimals, currency, and percent; and geometry and measurement. Mathematics Problem Solving samples single-step problems in the four basic operations and multistep problems (combined use of basic operations) involving currency, whole numbers, and fractions, decimals, and percents. Mathematics Computation samples computation skills with whole numbers, fractions, and decimals.

Selecting a Group Achievement Test

If teaching or assessment personnel become involved in the selection of a group achievement battery to be administered to all elementary (and middle school) children, content validity of the battery for the local school system should be of primary concern. All four batteries described possess excellent psychometric and standardization characteristics. All are well designed for their primary purposes: program evaluation, aids to formation of instructional groupings, and screening of high and low achievers. However, despite the common characteristics of standardized group achievement tests, no two tests are exactly alike. Each test measures somewhat unique aspects of content and skill (Gronlund, 1981) and the selection and sampling of mathematics objectives of the test battery must be matched with the objectives emphasized by the local curriculum.

Limitations of Group Norm-Referenced Tests

The mathematics objectives on group norm-referenced tests will be assessed quite thinly, often with only a single item. While many test publishers will provide reports on item performance for a "criterion-referenced" interpretation, these should be viewed with extreme caution. These performance reports might be suggestive of particular performance deficits, but would require further assessment for confirmation. Since most of the test items use multiple-choice formats, item performance will reflect the effects of guessing.

Finally, the performance of certain groups of learners will tend to be either overestimated or underestimated with respect to their true level of achievement. Learners who benefit from the opportunity to guess but whose performance on supply-type items would be extremely poor will tend to have overestimated test scores. Learners who do not respond in a motivated fashion under group testing conditions or who demonstrate reading disabilities will tend to test poorer than would be predicted from their actual level of mathematics achievement.

Criterion-Referenced Tests

The increased interest in recent years in criterion-referenced assessment has given impetus to a number of group criterion-referenced measures in mathematics. Some of these tests are independent of an instructional program, while others are closely linked to an instructional system, such as the Math Concept Inventory, the assessment component of *Project MATH* (Cawley, Goodstein, Fitzmaurice, Lepore, Sedlak, & Althaus, 1976). The Math Concept Inventory is reviewed in detail by Goodstein (1975). Several examples of those inventories not directly linked to an instructional system are briefly reviewed in the following section.

As indicated previously, the *Metropolitan Achievement Tests* now include a series of *Mathematics Instructional Tests*. The tests parallel the survey tests as to levels and mathematics strands included. The principal difference rests in the number of items (a minimum of three) to evaluate each objective. The *Mathematics Instructional Tests* are also norm-referenced. However, the design of the instrument allows for legitimate criterion-referenced interpretations.

• The *Diagnostic Mathematics Inventory* (DMI) (Gessel, 1975) is purportedly based on a comprehensive set of mathematics curriculum objectives from grades 1.5 through 8.5. It is divided into seven levels of one year each. Each level includes only those objectives ordinarily covered in that grade level. Some objectives are measured at more than one level (with the same item). Each concept (e.g., place value) is evaluated with a minimum of two items; the majority are evaluated with three or more items. Each item is keyed to a specific objective within the concept being evaluated. The scope of the DMI is comprehensive. However, the depth of the assessment is quite shallow. A *Guide to Ancillary Materials* and a *Learning Activities Guide* are available to the teacher for selecting or developing instructional materials correlated to DMI performance.

• The *Multilevel Academic Skill Inventory* (MASI) (Howell, Zucker, & Morehead, 1982) includes a mathematics component to evaluate basic skills objectives that range in curriculum level from grades 1 through 8. MASI is organized to assess at three levels. Survey tests provide a general sample of performance on key objectives. The results of the survey test suggest specific skill clusters to be assessed by placement tests (the second level of assessment). Finally, individual skills are evaluated by the specific level tests, where typically ten items are used to assess each skill. The MASI is divided between computation and application. Application skills include problem solving, money, time and temperature, metric measurement, customary measure, and geometry. The authors of MASI claim that suggested mastery criteria were derived from a review of academic skills mastery by normally achieving students and field tests with normal achievers.

• The *Individual Criterion-Referenced Tests* (Hambleton, 1981) consists of a series of criterion-referenced tests organized by strand (e.g., measurement, operations) within six levels intended for use in grades 1 through 6. The "Math Basics +" component evaluates 384 objectives. Levels are subdivided into booklets that contain eight objectives, with each objective evaluated by two items. Parallel forms of each booklet are available for retesting purposes. Performance is reported on an objective attainment basis (100 percent as mastery), as well as in reference to national norms.

GROUP ACHIEVEMENT TESTS: SECONDARY

As it becomes more apparent that many learners with mathematics disabilities in the elementary school continue to exhibit disability in the secondary grades, attention has begun to focus on assessment and remediation practices in the secondary school (Cawley, Fitzmaurice, Shaw, Kahn, & Bates, 1978). A key issue in assessment at this level is the functional level of the learner. For milder disability, where the learner is engaged in curriculum appropriate for other learners of his or her age, the preferred tactic would be to select instruments of appropriate content validity. For severe cases of disability, administration of tests designed to assess secondary level mathematics objectives is a fruitless exercise. Such tests have little content validity for these learners.

A preferred tactic is administration of a test that more closely parallels the functional level of mathematics achievement for the learner. Administering a level of a test designated for an age range that does not include the age of the learner to be assessed is referred to as "out-of-level" testing. This out-of-level testing with a norm-referenced test seriously compromises the ability of the examiner to use the normative data for interpretation (Berk, 1981). That is, since no learners of this older age were included in the normative population, the norms become inadequate for comparing the performance of the older learner. Of course, if the test to be given to the older learner is criterion-referenced or has a reasonable basis for a criterion-referenced interpretation of performance, out-of-level testing does not present problems.

Alternative Secondary Achievement Tests

Group achievement batteries are less widely used at the secondary level. Three alternative types of achievement tests are common at the secondary level (Gronlund, 1981). One type extends assessment of basic skills through the high school years. A second type emphasizes the course content and skills of the secondary mathematics curriculum. A third type (closely linked with aptitude testing) measures general educational development in intellectual skills and abilities that are not dependent upon specific courses (e.g., the *Iowa Tests of Educational Development*).

Examples of the first type of group achievement test include the *California Achievement Tests* (Level 19, grade levels 9.6–12.9), the *Metropolitan Achievement Tests* (Level Advanced 2, grade levels 10.0–12.9), and the *Stanford Achievement Tests* (TASK 2, grade levels 9.0–13.0).

- The *Sequential Tests of Educational Progress (STEP III)* (ETS, 1979), developed by the Educational Testing Service and commercially distributed by the Addison-Wesley Testing Service, is an example of the second type of achievement test. The target grade levels for the advanced level are from 10.5 to 12.9. In addition to tests of concepts and computation, algebra and geometry achievement tests are available.
- The *Iowa Tests of Achievement and Proficiency* (Scannell, 1978) might also be considered as a test that emphasizes secondary course content and is not an extension of an elementary battery. It is intended for grades 9 through 12. However, it also evaluates a number of "basic skills" objectives, whose grade level associated with mastery would fall considerably below grade 9.

Proficiency Tests

With the rapid adoption of proficiency tests to certify eligibility for high school graduation, commercial test publishers have attempted to capitalize on the need for tests that school systems might wish to administer to assess learner achievement of basic skills objectives. It would appear that in mathematics there is a greater (but by no means complete) level of agreement on a core set of basic skills. However, the usefulness of any commercial test to assist in directing remediation of basic skill deficits will depend on the content validity of the test with respect to each state's basic skills objectives.

The *Iowa Tests of Achievement and Proficiency* is partially an assessment of basic skills. The *Stanford Achievement Tests* (TASK 2) is a basic skills proficiency test. Scott-Foresman test division has recently introduced a *Minimum Essentials Test* based on those objectives considered basic to minimum competence in mathematics. Basic skills proficiency tests would appear to be an excellent assessment tool for initiating the evaluation of the secondary-age child with a mathematics disability.

INDIVIDUAL ACHIEVEMENT TESTS

While there exist many well-constructed group measures of mathematics achievement, norm- and criterion-referenced, administration of an individual achievement test is sometimes desirable. Special education mathematics assessments generally include individual rather than group measures (McLoughlin & Lewis, 1981). Individual achievement tests are often designed to minimize the effect of poor reading ability on mathematics achievement. The test materials are typically more attractive and less similar to classroom materials (worksheets and informal tests), with which the learner might associate previous failure. Of course, individual administration provides the opportunity for close observation of the learner and for motivation within the limitations of standardized administration procedures.

Norm-Referenced Tests

• The mathematics subtest of the *Peabody Individual Achievement Test* (PIAT) (Dunn & Markwardt, 1970) is one of the most popular individual achievement measures. The mathematics subtest consists of 84 multiple-choice items (four-option) to assess mathematics skills taught from kindergarten through grade 12. The test stresses application of mathematics skills. Concept areas include identification, matching, and naming numerals; basic whole number operations; fractions, decimals, and percentages; and measurement, geometry, algebra, and trigonometry.

The test items ultimately selected for the PIAT presented those psychometric properties that allowed for stable normative properties. Since no criterion-referenced interpretation of the PIAT was intended, it is somewhat unfair to criticize it for inadequate content coverage. However, it should be noted that there is probably little to be gained in the assessment of mathematics disability from administering the PIAT other than a comparison of the scores of the learner with the scores of the normative population.

Two limitations are noted by the authors. The content of the PIAT is limited to problems that can be solved mentally. Thus, problems that might require application of complex algorithms could not be included. Also, the items had to fit the multiple-choice format. Thus, no evaluation can be made of the ability of the learner to construct or supply the correct response.

• The *Woodcock-Johnson Psychoeducational Battery* (Woodcock, 1978) includes two subtests designed to assess mathematics achievement. The calculation subtest measures skill in solving paper-and-pencil addition, subtraction, multiplication, and division problems. The applied problems subtest assesses skill in solving practical arithmetic problems. These two subtests form a mathematics achievement cluster, the term *mathematics* suggesting a broader range of content than the arithmetic skills evaluated.

The Woodcock-Johnson is designed to be administered to persons from 3 to 80 years of age, although its main application will be with school-aged learners. The Woodcock-Johnson has been praised as a model for excellence in the development of norm-referenced psychoeducational test development (Salvia & Ysseldyke, 1981). However, its usefulness as a total battery appears greater than its usefulness as an individual achievement test in mathematics.

• The *Wide Range Achievement Test* (WRAT) (Jastak & Jastak, 1978) is divided into two levels. Level I is designed for use with children between the ages of 5 years 0 months and 11 years 11 months. Level II is intended for persons from 12 years 0 months to adulthood. The arithmetic subtest assesses counting, reading number symbols, solving oral problems, and performing written computations. Level I contains 43 computational problems that the student is directed to attempt to write answers for within a ten-minute time limit. Level II contains 46 computational problems ranging from simple arithmetic calculations to problems requiring knowledge of algebra and geometry. The WRAT will severely penalize those students who do not respond well to the speeded conditions of administration. The range of content is largely restricted to arithmetic computation skills.

Limitations of Individual Norm-Referenced Tests

Unfortunately, the three most commonly used individual achievement tests in mathematics were designed exclusively as norm-referenced instruments. Although these tests provide an estimate of a learner's performance with respect to the performance of a normative population, little diagnostic or prescriptive inference can be drawn from test results because of the extremely thin sampling of skills and concepts. McLoughlin and Lewis (1981) note that the arithmetic subtests of the WRAT and PIAT measure different skills. The WRAT arithmetic subtest requires a written response to computational problems, while the PIAT mathematics subtest requires a selection response to items involving not only computation, but reasoning and applications. They note the correspondingly low correlations between scores on the WRAT and the PIAT.

Criterion-Referenced Tests

As with group achievement tests, there is an increasing interest in supplanting or supplementing individually administered norm-referenced achievement tests with criterion-referenced measures.

• The *Brigance Diagnostic Inventory of Basic Skills* (Brigance, 1977) is intended for use with children functioning from kindergarten through sixth grade. The mathematics subscale includes 64 tests (sequences) organized into four sections: Numbers, Operations, Measurement, and Geometry. Each sequence is referenced to the grade level at which the content of the sequence is first taught. A norm-referenced computation test in the basic arithmetic operations is included with the Inventory.

• *System Fore* (Bagai & Bagai, 1979) assesses sequences of objectives from preschool through high school. Skill sequences are correlated with instructional objectives and arranged by developmental level from six months to 14 years. The mathematics sequences are arranged in three sections: Geometry and Measurement, Numbers and Numerals, and Operations and Applications. A feature of *System Fore* is the linkage of the instructional objectives with commercially available materials.

• *Criterion-Referenced Curriculum* (CRC) (Stephens, 1982) is the commercial revision of the assessment/instructional system outlined in *Teaching Children Basic Skills* (Stephens, Hartman, & Lucas, 1978). The CRC is referenced to the scope and sequence of skills in the primary grades. A total of 376 skill objectives in mathematics are referenced to the level at which handicapped children normally learn the skills in the elementary grades. Each assessment task is correlated with a set of instructional activities and materials. The CRC has been field tested over a ten-year period with mildly handicapped learners.

• *Diagnosis: An Instructional Aid in Mathematics, Level A* (Guzaitis, Carlin, & Juda, 1972) and *Level B* (Troutman, 1980) is designed to assess specific mathematics skills from kindergarten to third grade (Level A) and from third through eighth grade (Level B). Skills are grouped into six categories: whole numbers-concepts, whole numbers-addition and subtraction, whole numbers-multiplication and division, whole numbers-word problems, money and time, and geometry and measurement. Each skill is assessed by a criterion-referenced test (probe). A survey test is available to obtain a global assessment of the learner and to determine which specific probes to be administered. The assessment system is correlated with a wide range of elementary mathematics curricula.

Evaluating Criterion-Referenced Tests

Numerous other criterion-referenced inventories are also available, many of which have been developed locally by school systems. Criterion-referenced inventories should be evaluated with respect to various features. The items should be judged as to whether they are the best or most representative measurement tactic to assess the content suggested by the instructional objective. Sufficient items of similar content and format should be provided to reliably measure the objective. Items should be free of technical flaws. Objectives (and items) should be arranged in a reasonable developmental progression in order to facilitate efficient assessment. (Survey tests designed to aid in planning strategy for inventory administration may prove helpful.) Finally, the instructional objectives evaluated by the inventory should closely correlate with the instructional mathematics program for the child. Of course, if the assessment instrument is one component of a total curriculum system of instruction, content validity is ensured. Hambleton and Eignor (1978) provide an extensive listing of evaluative criteria.

Precision-Teaching Probes

While the term *probe* is often used synonymously with *criterion-referenced test* (e.g., as in the SRA *Diagnosis* systems), the original purpose of probes was as ancillary measurement systems for applied behavior analysis and precision-teaching systems. One of the distinguishing differences between probes and typical criterion-referenced tests is the provision of many more items to assess a specific skill. Howell, Kaplan, and O'Connell (1979) illustrate a sample addition probe of single-digit addition facts that includes 48 items. The large number of items is provided to allow for calculation of reliable rate data. The learner's performance on precision teaching probes is reported as percent correct or incorrect per minute. Thus, speed as well as accuracy significantly affects performance.

There may be mathematics skills that ought to be overlearned in order to establish automatic response patterns. The basic facts in whole number operations are examples of such skills. Precision teaching probes may prove to be useful in evaluating mastery of such computational skills. However, their usefulness is directly tied to the use of precision teaching or other applied behavior analysis remedial instruction systems. Moreover, for many mathematics skills, rate of responding would not be an appropriate metric. Thus, most precision teaching programs tend to be limited to learning computational or rote skills.

DIAGNOSTIC TESTS

If the school system typically places students requiring remedial assistance in mathematics into a criterion-referenced instructional program (managed through an assessment system), the assessment process may end with the administration of a criterion-referenced inventory or placement test. However, for learners with severe disabilities and for instructional planning in more "open" instructional systems, a diagnostic test in mathematics is often administered.

Definition of a Diagnostic Test

The presumed distinction between mathematics achievement tests and diagnostic tests resides in the intent of the diagnostic test to determine the cause of nonachievement (Goodstein, 1975). It should be pointed out that simply because a test author calls his or her instrument a diagnostic test does not necessarily endow it with diagnostic properties (Cronbach, 1970).

Gronlund (1981) suggests that diagnostic tests include more intensive measurement of specific skills, the use of test items that are based on a detailed analysis of specific skills involved in successful performance and a study of common errors, and the inclusion of items with lower levels of item difficulty than would be appropriate for survey achievement tests. The presumption being made is that a diagnostic achievement test would be more typically administered to learners with low levels of normative achievement. Underhill et al. (1980) suggest that diagnostic tests are more systematic over a narrower range of content.

Gronlund (1981) cautions that each diagnostic test reflects the author's concept of the subject area and viewpoint regarding diagnosis. Diagnostic tests do not necessarily indicate the cause of errors and differ widely in the ability to infer the cause of errors. While many experts in mathematics education have suggested that diagnosis not be limited to the abstract level and modeled after classroom situations, instrument development has been largely limited to symbolic skills. Unfortunately, clinicians and diagnosticians have not agreed upon standards for diagnostic instruments (Underhill et al., 1980).

Reisman (1982) summarizes the challenge of creating a comprehensive diagnostic mathematics test:

> It is impossible to construct a test that taps the total range of a mathematics curriculum and that provides information as to why a child misses one item and has answered the preceding item correctly. An analysis of published tests that purport to be diagnostic shows that between most items, several missing relations and concepts may exist. Furthermore, most tests do not provide enough information as to why a child has answered an item or a group of items incorrectly. (p. 44)

Norm-Referenced Diagnostic Tests

- The *Key Math Diagnostic Arithmetic Test* (Connolly, Nachtman, & Pritchett, 1976) is the most widely used diagnostic test of mathematics in special education (McLoughlin & Lewis, 1981). Key Math is an individually administered norm-referenced test designed to assess mathematics performance of learners from kindergarten through grade 8. The test is composed of 14 subtests organized into three areas: content, operations, and applications. Grade-equivalent scores are available for the total score and for individual subtest scores. Items are arranged within subtests in ascending order of difficulty and are calibrated (and displayed) with respect to the grade-equivalent scale.

Salvia and Ysseldyke (1981) observe that Key Math performance can be interpreted in four ways: total test performance, relative area strengths and weaknesses, relative performance on the 14 subtests, and a criterion-referenced interpretation based upon item performance and the description of behaviors sampled by each item. With respect to area strengths and weaknesses, it would appear that the analysis would be of dubious merit since no factorial validity has been demonstrated for the areas defined by Key Math (Goodstein, Kahn, & Cawley, 1976). With respect to a criterion-referenced interpretation of performance, this would appear to be limited by the fact that most objectives are measured by a single item. Substantial skill gaps exist between items. If a criterion-referenced evaluation of performance is desired, better assessment instruments are available.

The diagnostic value of Key Math is limited to comparisons of normative subtest performance. However, low subtest reliabilities for some subtests make such comparisons subject to considerable error (Salvia & Ysseldyke, 1981). Further, some subtests, such as Mental Computation and Missing Elements, may assess mathematics skills not found in the typical mathematics curriculum (McLoughlin & Lewis, 1981). The skill gaps between items and the inability to assess an objective with more than one item limits the interpretation of Key Math with respect to remedial strategies. Since it fails to intensively measure specific skills and provides no data with respect to causality for failure, Key Math more closely resembles a multiple-subtest individually administered achievement test than a diagnostic test.

Given the limitations of Key Math as a diagnostic mathematics test, it is difficult to understand its enthusiastic endorsement by authors of assessment textbooks. For example, Wallace and Larsen (1978) describe Key Math "as a good example of a comprehensive arithmetic battery that provides an overall indication of a child's arithmetic skills, along with more detailed information for teaching specific skills" (p. 449). Salvia and Ysseldyke (1981) proclaim its real value as a criterion-referenced device, with apparently little regard for the lack of comprehensive skill coverage. Perhaps, the endorsements reflect the current dearth of alternative individual diagnostic tests.

• The *Stanford Diagnostic Mathematics Test* (SDMT) (Beatty, Madden, Gardiner & Karlesen, 1976) is another widely used mathematics diagnostic test that is a group-administered measurement. The SDMT is organized into four levels: the Red level (grades 1.5 to 4.5), the Green level (grades 3.5 to 6.5), the Brown level (grades 5.5 to 8.5), and the Blue level (grades 7.5 to high school). At each level skills and knowledge are organized into three subtests: Number System and Numeration, Computation, and Applications.

The SDMT is essentially a norm-referenced instrument, with norm-referenced scores available for each subtest and total score. However, it contains many more easy items than do most achievement tests. Moreover, more items are provided for each skill area. This latter feature provides for criterion-referenced interpretations. Items that measure a specific skill are grouped and assigned a criterion score (called a Progress Indicator cutoff score). Cutoff scores were suggested by the authors judgmentally on the basis of the importance of the skill as a prerequisite for other skills and normative item difficulties. Patterns of performance or specific areas of weaknesses can be identified.

The SDMT is an exceptionally well-constructed group-administered measurement. Parallel forms exist that facilitate retesting for reevaluation of learner performance. As a group-administered measurement instrument, it may not be so motivating for the child as some of the more attractive individual tests (e.g., Key Math). It is more likely to yield instructionally relevant information on specific performance problems than Key Math. However, the SDMT may not contribute substantially in suggesting the cause of the poor performance.

Criterion-Referenced Tests

• Wallace and Larsen (1978) suggest that the *Diagnostic Chart for Fundamental Processes in Arithmetic* (Buswell & John, 1925) cannot be considered a test because of its lack of standardization, final scores, or quotients or grade equivalents. However, since it does present a standardized set of items to the learner in a standardized manner, it will be included in our discussion of diagnostic tests.

The learner is given a worksheet with a graded series of arithmetic computational problems arranged in order of increasing difficulty for each of the four basic operations with whole numbers. The learner is asked to work each of the problems aloud so as to assist the examiner in determining the algorithm employed in solving each problem. The need to consider the algorithm (or process) used that resulted in errors in computation is a consistent theme in the remedial mathematics literature (Goodstein, 1975; Underhill et al., 1980). The examiner classifies errors using the numerous categories provided on a diagnostic chart.

The Buswell-John instrument is basically a test of algorithms with no diagnosis regarding the structure of mathematics. No provision is made for diagnosing problem-solving techniques; and no provision is made for diagnosing place value either independently or in relation to algorithms (Underhill et al., 1980). However, the test is relatively easy to analyze for sequence of processes and learner errors are readily isolated.

• The *Sequential Assessment Mathematics Inventory* (SAMI) (Reisman, in press) should be included as a new criterion-referenced inventory. Described in *A Guide to the Diagnostic Teaching of Arithmetic* (Reisman, 1982), the SAMI is intended for use with learners from kindergarten through grade 8. The SAMI assesses mathematics content in 18 mathematics topics (ideas). Specific objectives (goals) are stated for each topic. Test items are classified by behavior (e.g., show, write, interpret, compute) required of the learner. The psychological nature (e.g., arbitrary association, relationship, concept, generalization) and mode of representing the content for each item (enactive, iconic, symbolic) are also specified.

Reisman describes the SAMI as a diagnostic screening instrument. Unlike typical criterion-referenced inventories, the domains represented by the topics (at each level of the instrument) are broader, inclusive of a number of related, but different, goals assessed with fewer items for each goal. Whether the ability to identify diagnostically meaningful patterns with respect to the various descriptive schemes for labeling items will contribute to effective remedial planning remains untested. The SAMI will constitute the most comprehensive single diagnostic instrument available. However, its interpretation will require considerable clinical experience and understanding of how children acquire mathematics knowledge and skill.

INFORMAL ASSESSMENT

Myers and Hammill (1982) contrast two assessment paradigms: the informal approach and the test-based approach. The informal approach relies heavily on the interpretation of children's performance in natural settings. It stresses criterion-referenced interpretations and is educationally task oriented. McLoughlin and Lewis (1981) include under the rubric of informal analysis work sample analyses (error analysis), checklists and inventories, criterion-referenced tests and inventories, and interviews and questionnaires. As commercially distributed criterion-referenced tests and inventories have been extensively reviewed earlier, this section will focus on the clinical interview, error analysis, and teacher-constructed tests and inventories.

Clinical Interview

The clinical mathematics interview is often suggested as an effective means of diagnosing the cause of performance deficits in mathematics (e.g., Cawley, 1978). Throughout the literature the interview technique emerges as a common thread in the various subjective methods of appraisal (Underhill et al., 1980).

Underhill et al. (1980) identified three approaches to the clinical interview: the structured interview, the Piagetian interview or clinique, and the teaching experiment. In the structured interview, questioning proceeds systematically in an attempt to assess student knowledge. All interviews are conducted in the same way and responses coded into predetermined categories and their relationships examined. The Piagetian interview begins in a structured manner but the examiner probes the responses made by the learner. No two interviews will be the same. Learner responses are recorded and transcribed and behavior patterns examined. The teaching experiment is used when the examiner is interested in determining how the learner acquires new knowledge. Questioning is used to assess learner understanding and to guide the learner.

Regardless of how the interview is structured, the effectiveness of the technique will depend upon the examiner's knowledge and ability to facilitate communication (Underhill et al., 1980). Clearly, effective use of the clinical interview would require significantly more training in mathematics education than is typically provided many assessment professionals (Wallace & Larsen, 1978).

Error Analysis

Error analysis is intended to determine the presence of systematic patterns underlying incorrect learner responses. Implicit in the use of error analysis is the expectation that many learner errors are not random or simply the result of carelessness. Rather, learner errors can provide diagnostic information to guide the remediation of specific concept or skill deficits (Goodstein, 1975).

Typically, three types of systematic errors may be identified: inadequate facts, incorrect operations, and ineffective strategies or algorithms (Cox, 1975; Moran, 1978; Reisman, 1982). Ashlock (1982) concentrates on defective algorithms and provides a wealth of examples for analysis.

Error analysis, which examines procedures, often yields data on the status of the learner that will stand in sharp contrast to an analysis based solely on scores. What might appear to be a total lack of comprehension may define itself as a single inappropriately applied concept or rule that quickly yields to precisely targeted instruction.

Error analysis can be made solely from the examination of the learner's work products. However, it is often combined with the clinical mathematics interview to enhance its diagnostic value (Cawley et al., 1978; Cox, 1975; Lepore, 1979).

Error analysis can also be effectively used in the informal assessment of verbal problem-solving errors (Goodstein, 1981). If verbal problems have been developed in such a manner that the various factors or parameters that combine to describe a problem have been controlled, organizational matrices can be developed (Goodstein, 1975). Such matrices allow for task descriptions that are sufficiently complete to evaluate whether error problems are random or systematic.

Teacher-Made Tests

Many authors have extolled the virtues of teacher-made tests in the assessment of mathematics disabilities (e.g., Bartel, 1978). The large number of criterion-referenced tests and inventories that are being commercially distributed would appear to obviate the need for teacher-made mathematics skill inventories. While the correspondence to the classroom curriculum may not be perfect, a far better strategy would be to modify or adapt a commercially distributed test or inventory. Objectives (and/or test items) could be added or deleted and additional parallel items could be developed to ensure greater reliability in the assessment process.

Teacher-made tests or inventories could prove quite helpful in providing more diagnostic information as to the nature and probable cause of specific skill deficits (Reisman, 1982; Underhill et al., 1980). These "informal" measures could assess the learner using a variety of formats, materials, and response modes that could not be cost-effectively employed by commercial test publishers. Unfortunately, the inability of teachers to prepare appropriate test items that will reliably assess learner performance is a major obstacle to valid teacher-made tests (Wallace & Larsen, 1978). Knowledge and understanding of the structure of mathematics, learning theories, developmental theories, and the nature of mathematics disabilities is essential.

Reisman (1982) offers the following guidelines for preparing teacher-made tests: (1) select content; (2) isolate one concept that is to be diagnosed in depth; (3) determine what level of learning the individual is at; (4) decide on what behaviors you want the learner to display in order to demonstrate acquisition of the concept; and (5) write a table of specifications that includes behavior and content components. Items would then be developed of sufficient technical quality to adequately sample the table of specifications.

A MODEL FOR ASSESSING SEVERE MATHEMATICS DISABILITIES

Risko (1981) describes three models commonly employed in the assessment of reading disabilities: the hierarchical plan, diagnostic-prescriptive assessment, and analytic teaching assessment. In the hierarchical plan three levels of assessment are used as needed. At the general level, achievement test data is analyzed in an effort to direct attention to the remediation of specific problem areas. At the analytical level, diagnostic tests are administered. At the case study level, a comprehensive evaluation is initiated. Each level of diagnosis is initiated only if remediation planning is unsatisfactory as a result of the assessment at the previous level.

Diagnostic-prescriptive assessment refers to criterion-referenced assessment within a skills-based instructional system. Analytic teaching assessment is based on applied behavior analysis (e.g., Howell et al., 1979). It would appear that these models are equally descriptive of assessment paradigms in mathematics disabilities.

The diagnostic-prescriptive assessment plan functions adequately only to the point where learners fail to sustain progress in the instructional system. The plan is predicated on the assumption that the skill sequence and instructional strategies optimize achievement. The learner who has a mathematics disability that has arisen as a result of unsystematic or inappropriate instruction most probably will respond well to diagnostic/prescriptive assessment and remediation.

For the learner who fails to sustain adequate progress, there are two alternatives: supplement or modify strategies (or sequence if possible) or change programs. In order to select and implement those choices, alternative assessment plans would have to be selected.

Analytic teaching assessment may be a useful adjunct to other assessment plans. However, its reliance upon observable (i.e., easily measured) responses severely limits the range of knowledge that can be assessed. Its inherent bias toward cognitive constructs restricts the consideration of alternative hypotheses regarding the cause of specific deficits.

Thus, the hierarchical plan in some form will be employed to assess the more severe levels of mathematics disability (e.g., Underhill et al., 1980). For it is at the analytical and case study levels that the most useful diagnostic information is likely to be developed to provide for effective remedial planning. It is at these levels that considerable research is being invested in the development of new diagnostic tests or inventories and the refinement of interview and observation techniques.

It is here also that nonmeasurement evaluations and clinical value judgments begin to have a major influence on the assessment process. Unfortunately, interpretation of the more analytic diagnostic tests and use of interview techniques require competencies omitted from many assessment training programs. This may yet prove to be the greatest impediment to progress in the assessment of mathematics disabilities.

REFERENCES

Ashlock, R.B. (1982). *Error patterns in computation: A semi-programmed approach.* Columbus, Ohio: Charles E. Merrill.

Bagai, E., & Bagai, E. (Eds.). (1979). *System Fore.* North Hollywood, Calif.: Foreworks.

Bartel, N.R. (1978). Problems in mathematics achievement. In D.D. Hammill & N.R. Bartel (Eds.), *Teaching children with learning and behavior problems* (pp. 99–146). Boston: Allyn & Bacon.

Beatty, L.S., Madden, R., Gardiner, E.F., & Karlesen, B. (1976). *Stanford Diagnostic Mathematics Test.* New York: Harcourt Brace Jovanovich.

Berk, R.A. (1981, May). *Identification of children with learning disabilities: A critical review of methodological issues.* Paper presented at the Johns Hopkins University Colloquium on Gifted/Learning Disabled Children, Baltimore, Md.

Boehm, A.E. (1971). *Boehm Test of Basic Concepts.* New York: Psychological Corporation.

Brigance, A. (1977). *Brigance Diagnostic Inventory of Basic Skills.* North Billerica, Mass.: Curriculum Associates.

Brigance, A. (1978). *Brigance Diagnostic Inventory of Early Development.* North Billerica, Mass.: Curriculum Associates.

Bruner, J. (1960). *The process of education.* Cambridge: Harvard University Press.

Buswell, G.T., & John, L. (1925). *Diagnostic Chart for Fundamental Processes in Arithmetic.* Indianapolis: Bobbs-Merrill.

Cawley, J.F. (1978). An instructional design in mathematics. In L. Goodman & J.L. Wiederholt (Eds.), *Teaching the learning-disabled adolescent.* Boston: Houghton Mifflin.

Cawley, J.F., Cawley, L.J., Cherkes, M., & Fitzmaurice, A.M. (1980). *Beginning Educational Assessment.* Glenview, Ill.: Scott, Foresman.

Cawley, J.F., Fitzmaurice, A.M., Shaw, R.A., Kahn, H., & Bates, H. (1978). Mathematics and learning disabled youth: The upper grades. *Learning Disability Quarterly, 1*(4), 37–52.

Cawley, J.F., Fitzmaurice, A.M., Shaw, R.A., Kahn, H., & Bates, H. (1979). LD youth and mathematics: A review of characteristics. *Learning Disability Quarterly, 2*(1), 29–44.

Cawley, J.F., Goodstein, H.A., Fitzmaurice, A.M., Lepore, A.V., Sedlak, R., & Althaus, V. (1976). *Project MATH: Levels I–IV*. Tulsa, Okla.: Educational Progress Corporation.

Connolly, A.J., Nachtman, W., & Pritchett, E.M. (1976). *Key Math Diagnostic Arithmetic Test*. Circle Pines, Minn.: American Guidance Service.

Cox, L.S. (1975). Diagnosing and remediating standard errors in addition and subtraction computations. *The Arithmetic Teacher, 22*, 151–157.

Cronbach, L.J. (1970). *Essentials of psychological testing*. New York: Harper & Row.

CTB/McGraw-Hill. (1977). *California Achievement Tests*. Monterey, Calif.: Author.

Dunn, L.M., & Markwardt, F.C. (1970). *Peabody Individual Achievement Test*. Circle Pines, Minn.: American Guidance Service.

ETS. (1979). *Sequential Tests of Educational Progress*. Reading, Mass.: Addison-Wesley.

Gagné, R.M. (1977). *Conditions of learning* (3rd ed.). New York: Holt, Rinehart & Winston.

Gardiner, E.F., Rudman, H.C., Karlesen, B., & Merewin, J.C. (1981). *Stanford Achievement Tests*, New York: Psychological Corporation.

Gessel, J.K. (1975). *Diagnostic Mathematics Inventory*. Monterey, Calif.: CTB/McGraw-Hill.

Glennon, V.J., & Wilson, J.W. (1972). Diagnostic-prescriptive teaching. In W.C. Lowrey (Ed.), *The slow learner in mathematics* (pp. 282–318). (Thirty-fifth Yearbook of the National Council of Teachers of Mathematics.) Washington, D.C.: NCTM.

Goodstein, H.A. (1975). Assessment and programming in mathematics for the handicapped. *Focus on Exceptional Children, 7*(7), 1–11.

Goodstein, H.A. (1981). Are the errors we see the true errors?: Error analysis in verbal problem solving. *Topics in Learning and Learning Disabilities, 1*(3), 31–46.

Goodstein, H.A. (1982). The reliability of criterion-referenced tests and special education: Assumed versus demonstrated. *Journal of Special Education, 16*, 37–48.

Goodstein, H.A., Kahn, H., & Cawley, J.F. (1976). The achievement of educable mentally retarded children on the Key Math Diagnostic Arithmetic Test. *Journal of Special Education, 10*, 61–70.

Gronlund, N.L. (1981). *Measurement and evaluation in teaching*. New York: Macmillan.

Guzaitis, J., Carlin, J.A., & Juda, S. (1972). *Diagnosis: An Instructional Aid (Mathematics), Level A*. Chicago: Science Research Associates.

Hambleton, R.K. (1981). *Individual Criterion-Referenced Testing: Math Basic +*. Tulsa, Okla.: Educational Development Corporation.

Hambleton, R.K. & Eignor, D.R. (1978). Guidelines for evaluating criterion-referenced tests and test manuals. *Journal of Educational Measurement, 15*, 321–327.

Hieronymous, A.N., Lindquist, E.F., & Hoover, H.D. (1979). *Iowa Test of Basic Skills*. Lombard, Ill.: Riverside.

Howell, K.W., Kaplan, J.S., & O'Connell, C.Y. (1979). *Evaluating exceptional children: A task analysis approach*. Columbus, Ohio: Charles E. Merrill.

Howell, K.W., Zucker, S.H., & Morehead, M.K. (1982). *MASI: Multilevel Academic Skill Inventory*. Columbus, Ohio: Charles E. Merrill.

Jastak, J.F. & Jastak, S. (1978). *Wide Range Achievement Test*. Wilmington, Del.: Jastak Associates.

Jensen, A.R. (1980). *Bias in mental testing*. New York: The Free Press.

Johnson, S.W. (1979). *Arithmetic and learning disabilities*. Boston: Allyn & Bacon.

Kagan, N. (1976). *Interpersonal process recall—elements of facilitating communication, Parts I and II*. East Lansing: Michigan State University.

Kamii, M. (1981). Children's ideas about written number. *Topics in Learning and Learning Disorders, 1*(3), 47–60.

Kirk, S.A., McCarthy, J.J., & Kirk, W.D. (1968). *Illinois Test of Psycholinguistic Abilities* (rev. ed.). Urbana, Ill.: University of Illinois Press.

Lepore, A.V. (1979). A comparison of computational errors between educable mentally handicapped and learning disability children. *Focus on Learning Problems in Mathematics, 1*, 12–33.

Lovitt, T.C. (1975). Applied behavior analysis and learning disabilities: Part II. Specific research recommendations and suggestions for practitioners. *Journal of Learning Disabilities, 8*, 36–50.

Madden, R., Gardiner, E.F., & Collins, C.S. (1981). *Stanford Early School Achievement Test*. New York: Psychological Corporation.

McEntire, E. (1981). Learning disabilities and mathematics. *Topics in Learning and Learning Disabilities, 1*(3), 1–18.

McLoughlin, J.A., & Lewis, R.B. (1981). *Assessing special students*. Columbus, Ohio: Charles E. Merrill.

Moran, M. (1978). *Assessment of the exceptional learner in the regular classroom*. Denver: Love Publishing.

Moss, M.H. (1978). *Tests of Basic Experiences 2*. Monterey, Calif.: CTB/McGraw-Hill.

Myers, P.I., & Hammill, D.D. (1982). *Learning disabilities: Basic concepts, assessment practices and instructional strategies*. Austin, Tex.: Pro-Ed.

Nurse, J.R., & McGauvran, M.E. (1976). *Metropolitan Readiness Tests*. New York: Harcourt Brace Jovanovich.

Piaget, J. (1932). *The language and thought of the child*. New York: Harcourt Brace Jovanovich.

Prescott, G.A., Balow, I.H., Hogan, T.P., & Farr, R.C. (1978). *Metropolitan Achievement Tests: Survey Battery and Instructional*. New York: Psychological Corporation.

Reid, D.K., & Hresko, W.P. (1981). *A cognitive approach to learning disabilities*. New York: McGraw-Hill.

Reisman, F.K. (1982). *A guide to the diagnostic teaching of arithmetic*. Columbus, Ohio: Charles E. Merrill.

Reisman, F.K. (in press). *Sequential Assessment in Mathematics Instruction (SAMI)*. Columbus, Ohio: Charles E. Merrill.

Risko, V.J. (1981). Reading. In D. Deutsch-Smith, *Teaching the learning disabled*. Englewood Cliffs, N.J.: Prentice-Hall.

Salvia, J., & Ysseldyke, J.E. (1981). *Assessment in special and remedial education*. Boston: Houghton Mifflin.

Scannell, D.P. (1978). *Iowa Tests of Achievement and Proficiency*. Lombard, Ill.: Riverside.

Stephens, T.M. (1982). *CRC: Criterion-Referenced Curriculum*. Columbus, Ohio: Charles E. Merrill.

Stephens, T.M., Hartman, A.C., & Lucas, V.H. (1978). *Teaching children basic skills: A curriculum handbook*. Columbus, Ohio: Charles E. Merrill.

Troutman, A.P. (1980). *Diagnosis: An Instructional Aid (Mathematics), Level B*. Chicago: Science Research Associates.

Underhill, R.G., Uprichard, A.E., & Heddens, J.W. (1980). *Diagnosing mathematics disabilities*. Columbus, Ohio: Charles E. Merrill.

Wallace, G., & Larsen, S.C. (1978). *Educational assessment of learning problems: Testing for teaching*. Boston: Houghton Mifflin.

Wick, J.W., & Smith, J.K. (1980). *Comprehensive Assessment Program: Achievement Series*. Glenview, Ill.: Scott, Foresman.

Woodcock, R. (1978). *Woodcock-Johnson Psychoeducational Battery*. Boston: Teaching Resources.

Chapter 4

An Integrative Approach to Needs of Learning-Disabled Children: Expanded Use of Mathematics

John F. Cawley

Children referred to as learning disabled are generally viewed as a heterogeneous group. In fact, the heterogeneity of characteristics subsumed under the concept of learning disability is so great that agreement as to what is truly characteristic of the learning disabled is near impossible. The phenomenon of heterogeneity creates numerous problems. These involve

1. the identification of those specific characteristics that will help the field to identify needs and subsequently develop effective programs
2. the identification or development of instructional activities or curricula that will modify the characteristics
3. our ability to modify a given set of characteristics (i.e., visual perceptual characteristics) to the betterment of another set of characteristics (e.g., addition computation, equivalence classes of visually presented fractions)
4. our ability to isolate characteristics and to focus upon one set of needs to the exclusion of others (e.g., provide tutorial experiences in decoding independent of experiences with numbers)
5. the possibility that we have failed to develop an integrative approach in which the strengths or weaknesses in one or more sets of attributes impact or compensate for needs in others
6. the possibility that exclusion of natural experiences from the educational programs of the learning disabled because they are "not relevant" or because they are "too difficult" may be a detrimental educational decision.

Dependency upon mathematics extends to numerous activities (e.g., knowing the time to go to recess, understanding subject-verb relationships in language, knowing the location of one's chair, giving only part when "some" is asked for). In fact, the stimulus properties and qualities of mathematics are such that activities rooted in mathematics can be integrated into the full range of curriculum options in both subject matter (e.g., science, social studies) and skill areas (e.g., reading, language). That any given set of needs of learning-disabled children can be focused on during subject matter instruction is a solid basis upon which to construct program options. In some instances these might provide a direct remedial emphasis, whereas in others they may be used to reinforce or to generalize a given concept or skill.

Process-type activities represent a legitimate component of the educational experiences for the learning disabled. For example, it seems as appropriate to ask a child the divergent-thinking (process) question "How many ways can you write 6?" as it is to ask a convergent-thinking or cognitive-memory question, "Will you write the answer to 4 plus 2?" The former represents a generalization of the latter, assuming that the latter is understood as a concept and not simply a response to rote learning.

Limiting our efforts to extensive computational practice to learn the "addition facts" by either paper-pencil or computer-based instruction is misguided. Better to teach the child a strategy to learn two or three combinations (e.g., $3 + 2 = 5$; $4 + 2 = 6$; $6 + 3 = 9$) and then hold the child responsible for using the strategy to learn others as they are encountered or assigned. The paper-pencil or computer-based activities can then be used to enhance speed and facilitate habituation.

Processes such as decision making, creating alternatives, generalization, and application should be integral components of the program. And, these processes should be integrated into as many activities as possible. Thus, the integrative approach may be defined as an approach that integrates a variety of strengths or weaknesses into an individualized educational program that acknowledges that a child attends school all day long.

THE INTEGRATIVE APPROACH: WORD PROBLEMS

Although many different emphases could be developed, we will focus on the role that word-problem activities can play in programs for learning-disabled children and how word-problem activities can be used to address selected needs through an integrative approach. The premise is that word-problem activities can do much to facilitate many aspects of growth and development. Because of this, they should be considered lead items in programs rather than activities to which children are exposed after they learn to compute.

Word-problem activities are situational activities in which words, the structures in which the words are arranged, and the contexts in which the words are embedded control the actions and decisions of the child. These may consist of activities that focus upon specific types of thinking such as *class inclusion* (Hierbert, Carpenter, & Moser, 1982) and *comparisons* (Feuerstein, 1980) or any other factors that require the individual to process problems beyond the levels at which they have become so habituated that they are no longer problems (Shaw, 1981). Tasks that direct the child to seek key words (i.e., left means to subtract) or to use computational rules (e.g., three or more different numerals suggest addition) as aids to the selection of an operation are not word-problem activities. They are, in reality, instances of negative learning for the child.

In an integrative approach two or more basic areas of curriculum and instruction are interrelated through the use of a common theme or topic. Specific sets of needs in areas such as reading, language, and mathematics are identified and the relevant instructional activities are incorporated into the theme. Once the theme has been selected and the needs of the learner established, the next step is to arrange the instructional program into a progression. This progression will start with the most basic needs (e.g., reading vocabulary) and progress through to the set of target needs (e.g., interpreting language statements).

The integrative approach reduces the need for the child to learn many new things at one time or to have to utilize many diverse skills or sets of knowledge at any one time.

The illustrations developed in this chapter are related to the following characteristics:

1. deficiencies in reading in decoding and/or comprehension
2. deficiencies in language in reception and/or expression
3. deficiencies in mathematics in computation and/or problem solving
4. deficiencies as participants in classroom activities
5. deficiencies in attention
6. deficiencies in memory or remembering

These few characteristics have been selected to illustrate some of the qualities of an integrative approach as a component of an appropriate educational curriculum for learning-disabled children. The numerous other characteristics and needs of these children, both positive and negative, can be incorporated into the integrative approach by adapting the examples described below.

BEGINNING THE SCHOOL DAY: REINFORCING READING AND COMPREHENSION

Examine the following:

Catching Fish

A big black bear is fishing in the lake. He is standing near the shore where the lake is not deep. The bear is smacking his lips. He is thinking how the fish will taste when he eats them.

> The bear is catching more fish than he can eat. When he is full of fish, he will leave the rest on the shore. A fox or a mink will come and eat them.

This story, which is part of a series of reading lessons (McCracken & Wolcutt, 1968) on which the children had been working, is designed to provide practice in context of the use of *ing* and reinforcement of the rules underlying *ing*.

The day begins with an activity in which one child goes to the chalkboard and writes a word (i.e., *jump*) and another child goes to the board and adds *ing*. A number of replications are undertaken and when the children demonstrate that they have acquired the *ing* idea, they are provided with a list of *ing* words to copy, read, act out, and discuss. The children are then introduced to stories that contain *ing* words where there is also an emphasis on comprehension. The teacher's edition stipulates that the following questions be asked:

1. What is the bear doing with his lips?
2. What is he thinking about?
3. What will happen to all the fish that will be left on the shore?

The reading lessons have taught a skill (i.e., *ing*), developed vocabulary, and provided experiences with reading in context and comprehension. The teacher plans to capitalize on the motivating qualities of the topic and interests of the learners to extend the reading activities into the realm of mathematics. (Although the present illustrations are related to the topic of bears, any topic could be used in a similar manner.)

CONTINUING THE SCHOOL DAY: ON TO MATH

The children have moved on to mathematics and the teacher has rewritten the story in the following manner:

> Catching Fish
>
> Two black bears are fishing in the lake. One bear is standing near the shore. This bear has already caught 5 fish. The other bear is sitting in the water holding on to the 6 fish that she caught.
>
> Each bear is thinking about catching some more fish. Each bear is also thinking about eating the fish. After each bear caught 2 more fish, they went home.

This second story continues the reinforcement of *ing*. In fact, with fewer words, it provides more instances of *ing* than the original story. But this story is different in that it contains a number of quantifiers, thereby opening the door for an alternative set of questions, such as the following:

1. How many bears were thinking about catching some more fish?
2. How many fish were caught altogether by the bear sitting in the water?
3. How many fish did the bears catch in all?
4. Which bear caught the most fish?
5. What two things were the bears thinking about?
6. Why do you think the bears left after catching only two more fish?

Questions 1 through 4 direct the attention of the child to quantitative responses, question 5 is directed toward specific information, and question 6 is inferential.

Let us assume the teacher wanted to expand the story to extend the quantitative elements. The following might be considered:

1. increase the number of bears
2. increase the number of fish
3. introduce other animals (e.g. fox/mink)
4. vary the type of fish (e.g. trout/catfish)
5. vary the type of bears (e.g. brown/black)
6. vary the positions of the bears
7. increase the number and variety of questions

If the teacher wanted to vary the problem activity and focus on problems of the extraneous information type, the following might be considered:

1. A black bear standing in the water caught 4 fish. Another black bear standing in the water caught 3 fish. A brown bear standing in the water caught 2 fish. How many fish did the black bears catch?
2. A black bear standing in the water caught 4 fish. Another black bear standing in the water caught 3 fish. A black bear sitting in the water caught 2 fish. How many fish did the bears standing in the water catch?

In the first instance, the teacher used the adjective *brown* as the source of the extraneous information and in the second instance the verb *sitting* was used.

Upon noticing that the children were doing an excellent job with problems of the extraneous information type, the teacher decided to introduce indirect problems such as, "The black bear had 5 fish left after giving 2 fish to a fox. The brown bear had 2 fish left after giving 6 fish to a fox."

1. Which bear had the greater number of fish to begin with?
2. How many fish did the black bear start with?

3. How many fish were given away by the bears?
4. Which bear gave away the most fish?
5. How many fish did the bears have at the beginning?
6. How many fish did the brown bear keep?

CONTINUING THE SCHOOL DAY: ON TO LANGUAGE ARTS

Today's language arts objectives focus on the development of synonyms and antonyms and in the use of inflection and intonation to give meaning.

The teacher begins the activity by writing on the chalkboard, "The black bear thought that having 3 fish was great. The brown bear thought that having 2 fish was super." The children read each sentence as a choral reading activity. The teacher then reads each sentence aloud and asks, "What kinds of feeling or mood do the sentences tell us about the bears?" After some discussion as to how these two different words convey a similar meaning, the teacher asks, "Why do you think the brown bear thought that having two fish was super when the black bear had three fish?"

Children might respond with suggestions to the effect that (1) the brown bear had bigger fish—an indication that she likes fish, (2) the brown bear thought that having two fish was super because she had only two fish—an indication that she did not like fish, or (3) the fish were of a different and preferred type.

The teacher highlights the similarities between *super* and *great* and writes, "The black bear thought that having 3 fish was great. The brown bear thought that having 2 fish was terrible."

The teacher asks the children to signal the differences in the two sentences. The response focuses on *great* and *terrible* and the teacher elicits discussion as to their meanings.

Next the teacher asks, "Why do you think the brown bear felt terrible?" In all likelihood the responses focus on the quantifiers 3 and 2. The fact that the black bear had one greater can be contrasted with the fact that the brown bear had one fewer. In this instance, factors such as size, quality, and type of fish are likely to be overridden by the greater/fewer contrast.

As a final activity, the teacher writes on the board, "The black bear thought that having 3 fish was awful. The brown bear thought that having 2 fish was terrible." A contrast in the sentences highlights the feelings engendered by *terrible* and *awful*. What factors make having 3 fish or having 2 fish a less than satisfactory arrangement? Could it be that other bears had a greater number of fish? Could it be that the black bear and the brown bear were selfish?

Throughout the comparisons of the sentences, the feelings of the bears were mediated by the quantifiers 3 and 2. Note that no direct computation was involved. Thus, the verbal behaviors exemplified by the synonyms and antonyms and the feelings of the bears were mediated without extensive computation and without a need for the children to have common answers.

Let us assume the teacher wanted to expand upon the vocabulary components of the activity. The teacher writes, "The black bear thought that having 3 fish was ______. The brown bear thought that having 2 fish was ______." The teacher directs each child to produce *as many words as possible* to complete each sentence. By doing this the teacher develops a vocabulary list of words known to various, and perhaps all, members of the class. These words can be used to expand the vocabulary of the class and to assist children to discover the common meanings of different words and the different meanings of the same words.

When the children have "caught on" the teacher can decide to vary the activity by writing the following:

The black bear thought that having ___ fish was great.
The brown bear thought that having ___ fish was super.
The black bear thought that having ___ fish was great.
The brown bear thought that having ___ fish was terrible.

Each child could be instructed to put in the numbers he or she thinks would give meaning to the sentences. One child might insert 1 in each instance, in the first pair of sentences, a factor that would equate greater/super numerically. Another child might insert 12 in each of the first two sentences, giving numerical equality to the terms, but expressing satisfaction with larger numbers than the other child. Comparisons and discussion could follow.

If various children inserted the same or different numbers in the great/terrible sentences, elaboration, comparison, and analysis could be undertaken.

SCHOOL: ANOTHER DAY AND TIME FOR MATHEMATICS

Today's lesson will focus on the use of indefinite quantifiers as a component of the problem-solving process. Indefinite quantifiers are terms such as *some, many, few, bunch, group,* and *set.* The terms do not express cardinality. Rather, they direct the learner to attend to the "how manyness" of the quantity. Examine the pictures of the bear. Pictorial representations (see Figure 4–1) and manipulative activities are encouraged in mathematics programs with the learning disabled. However, in order for the pictures or the manipulatives to be effective, the attention of the learner must be directed toward them.

If one said, "Listen. The bear that was bitten by the bees has 3 fish and the bear that is biting the bees has 3 fish. How many fish do these bears have in all?" it is unlikely that the children would look at the pictures because all the needed information has been transmitted auditorially.

Figure 4–1 Pictorial Representation of Lesson

If the statement was changed to "The bear who was bitten by the bees has some fish and the bear who is biting the bees has some fish. How many fish do these bears have in all?" the learner would be required to look at the picture to determine how many fish each bear had. The indefinite quantifiers require the child to examine the pictures or the manipulative activity and to attend to them as sources of needed information. Once the youngsters have had their attention focused on the pictures, the teacher can then incorporate tense and voice as parts of the activity.

Variations in tense could be undertaken by preparing statements such as:

1. The bear being bitten by the bees has some fish. How many fish does this bear have?
2. The bear biting the bees has some fish. How many fish does this bear have?
3. One bear is being bitten by the bees. Another bear has been bitten by the bees. How many more fish does one have than the other?

Variations in voice can be undertaken by preparing statements such as:

> The bear biting the bees has some fish.
> The bees are being bitten by a bear who has some fish.
> Find the picture for each sentence.
> How many fish in all?
> How many bears?
> How many bees?

The answer will always be, 1 bear, 3 fish, and 2 bees, although some children may not see that the variations are permutated on the basis of active or passive voice and not upon a representation of differing numbers of animals. The inclusion of the indefinite quantifier aids in focusing attention on the correct choices.

SCHOOL: ON TO LANGUAGE ARTS

The focus of this lesson is on modification, in particular the use of a prepositional phrase to modify an object.

The teacher reads these statements, "The black bear gave the 3 trout behind the tree to the fox. The black bear gave the 2 catfish behind the tree to the fox." Based on the information in the two statements, the children must indicate TRUE, FALSE, or CAN'T TELL for the following:

1. The black bear was behind the tree.
2. The catfish were behind the tree.
3. The tree was behind the fox.
4. There was only 1 fox.
5. The black bear gave away 5 fish.
6. The black bear gave away more catfish than trout.
7. The black bear gave away more trout than fish.

One of the more important components of this lesson is the "Can't Tell" condition. It will be described below as the basis of activities the teacher can use to address some of the problems of the nonparticipating child.

DEVELOPMENTAL PROBLEMS

The illustrations have shown how it is possible to develop word-problem activities in an integrative manner. Reading, mathematics, and language arts activities using a common theme were integrated to provide reinforcement and to introduce new concepts and ideas within a set of common knowledge. Let us now turn our attention to the use of problem activities as an approach to developmental problems.

Participants/Nonparticipants

It is easy to distinguish between the child who sits on the edge of his or her seat making every effort to absorb every possible bit of knowledge, who eagerly awaits every opportunity to answer questions, offer alternatives, and become a member of every activity group, and who analyzes and differentiates among all types of stimuli and the child who performs in a perfunctory manner with only a minimal response and no sense of involvement.

What can we do about the nonparticipating child? Two factors have come into play. One is fear of failure or the anticipation of failure. The second is a genuine feeling of lack of self-esteem or self-worth.

Some children avoid participation because it seems as though everything they do is wrong. They develop response sets such as "I don't know" or "I forgot" or "I didn't do it." The "I don't know" response may not mean that the child does not know. It could mean "I am not certain and I don't want to take a chance on failure" or "I really don't want to play the game." Such a child may truly believe that he or she cannot learn and may even become terribly confused or ineffective even when dealing with simple tasks. This child needs to be successful, not just rewarded, over long periods of time and to develop an awareness of the meaning of success. In dealing with this child we might:

1. Structure the tasks to yield a high probability that the correct response will occur (e.g., change the extraneous information problem about the black and brown bears to a multiple-choice format in which the choices are 7 or 5. This clearly indicates that 10 is not an appropriate response and that the option to add all three numbers is minimized.
2. Arrange questions and other means of eliciting responses so that a number of questions follow each information set (e.g., the compound-indirect example suggests that question 6 might be more appropriate for the failure-prone child because it requires only that *left* and *keep* be understood as representing a comparable meaning). Different questions can be answered by different children, the net effect being a form of peer tutorial.

The child who lacks self-esteem or a feeling of self-worth believes that his or her ideas, contributions, and roles have no meaning or value to others. Not only is it important to provide this child with an opportunity to be of value, it is also necessary to help the child see the value of an activated or heightened state of participation. To approach this child one might:

1. Identify one or two of his or her interests and create word-problem activities using these interests as the theme. It is easy to substitute red cars, green cars, and a parking lot for brown bears, black bears, and a lake.
2. Actually take the children fishing. Have this child serve as an organizer.
3. Conduct fishing type activities such as bobbing for apples. These can be team activities. Scores can be kept and quantitative comparisons made.
4. Have the children create three-dimensional representations in the form of a diorama and prepare models of the bears, fish, and so forth and use them as sources of information in word problems. This type of activity modifies the ambiguity of the information in the True, False, Can't Tell conditions. By locating the bears and other objects in selected positions, the "Can't Tell" statement (e.g., "The black bear is behind the tree") is reclassified to either a True or False condition. The diorama might be constructed on a grid and the teacher could designate the placement of objects as points on a coordinate plane (e.g., "Put the bear at 4, 7").
5. Assign the child to key roles in a play or an enactment of the theme of the word-problem activity.
6. Arrange the activity so that records are kept. Give the child a record-keeping role.
7. Have the child assume the role of the teacher. Require that the stimuli given to the class be offered by the target child and that he or she be responsible for its interpretation to the group.
8. Have the child serve as the source of the answers. Work with the child to develop a set of correct answers and have other children consult this child for the correct answers.

Attentional Deficits

Attentional deficits tend to be of two types. In one instance, the child does not focus on the relevant dimensions of the activity or materials. If you are discussing bears, this child neglects to attend to factors such as size, color, position, and so forth. The second type of attentional deficit relates more to paying attention in the traditional sense. Here the child daydreams, strays off task, or focuses on other stimuli. The first child might be paying attention but not attending to what is relevant. The second child might be perfectly capable of attending to the relevant material if one could get him or her to pay attention.

Through the use of a story or pictorial representation of story content we can direct the attention of the child to relevant components of the story. This might entail activities such as the following:

1. Asking the child to draw a picture showing the relative locations and positions of the two bears.

2. Asking the child to draw a picture of a third bear and to show the location and position of this bear.
3. Asking the child to list characteristics of the bears and fish as shown in the pictures.
4. Indicating that certain conditions will be the focus of the follow-up (e.g., the color of the different bears in the extraneous information activities) and that the child will be held responsible for a specific task in relation to this (e.g., will be given a picture of a bear and be expected to color the picture to match the information given in the story).

The child who wanders off task might be approached with activities such as the following:

1. The children are given a key word in the story (e.g., *fish*). Each time the child hears the word as the teacher reads the problem aloud, the child must raise his or her hand. The teacher can glance about the room and determine the extent to which the target child is on task.
2. The teacher indicates that each child is to read the story and then answer the questions. Each child will be timed (e.g., story is on one side of paper. Child reads story, signals when finished, records time, and turns page over to questions. Child is timed again upon completion of the questions. Number of correct answers is contrasted with time. Number correct must be high). Teacher indicates that children will be asked to read story aloud and that at specific intervals another child will be asked to continue the reading. Target child may be asked to read first sentence and teacher may return to this child for a later sentence.

Memory Deficits

Memory deficits involve the inability to recall information that was recently presented (e.g., within a minute or two) or to remember information that was presented some time back (e.g., yesterday, two weeks ago). A key factor in memory is acquisition. That is, if the child did not learn the information in the first place, then it is difficult to assess memory. For example, in a study of word problems, Goodstein and Sedlak (1974) found that learning-disabled children required more repetitions of orally presented word problems before they could restate the problem as given than did average or *mildly retarded* children. In effect, it took the learning-disabled children more repetitions to remember the information. However, when asked to solve the problem after the information had been remembered, these children performed the problems correctly more often than the mildly retarded children who had acquired the information sooner.

The effects of memory deficits can be minimized by:

1. reducing the amount of material the child is expected to remember
2. repeating the material more often
3. developing experiences with the concepts and content of the information so as to extend its meaningfulness

Referring back to the extraneous information activities, we might modify the information set from "A black bear standing in the water caught 4 fish" to "A black bear caught 4 fish" or from "A black bear standing in the water caught 3 fish" to "A bear in the water caught 4 fish."

Pictures in language arts activities could be shown to the child with the indication that they will be taken away after a period of time and that the child will be required to use the information in a subsequent task.

WHERE TO GO FROM HERE?

Factors such as age, interests, level of ability, present level of functioning, and many others will serve to guide the selection of the most appropriate activities for the learning disabled. Special education teachers could use themes from the regular class and incorporate them into the instructional program. Regular education teachers could incorporate specific needs into the ongoing themes. The result should be a more wholistic approach to the needs of the learning disabled.

Certainly, there are many more word-problem activities and aspects of such activities. And certainly there are many more characteristics of children with learning disabilities that require attention. The task of providing meaningful and beneficial problem activities for children as heterogeneous as this target population is challenging and demanding.

Handicapped children need problem-based activities as much as nonhandicapped children. They need them on a continuous and daily basis throughout their school years in order that they can become problem solvers and users of information and concepts in the decision-making process. It is hoped that someday problem-solving activities will supersede computation as the primary source of mathematical experiences and that this emphasis will be found in geometry, measurement, fractions, and other mathematical strands. When this is accomplished, learning-disabled children will have a reason for wanting to learn to compute. And their teachers will have a reason to teach them to compute.

REFERENCES

Feuerstein, R. (1980). *Instrumental enrichment: An intervention program for cognitive modifiability.* Baltimore: University Park Press.

Goodstein, H.A., & Sedlak, R. (1974). The role of memory in the verbal problem solving of average, educable mentally retarded and learning disability children. Unpublished manuscript, University of Connecticut, Storrs.

Hierbert, J., Carpenter, T., & Moser, J. (1982). Cognitive development and children's solutions to verbal arithmetic problems. *Journal for Research in Mathematics Education, 13,* 83–98.

McCracken, G., & Wolcutt, C. (1968). *Basic reading* (Teacher Ed.). Philadelphia: J.B. Lippincott.

National Council of Teachers of Mathematics. (1980). *Priorities in school mathematics.* Reston, Va. Author.

Shaw, R. (1981). Designing and using non-word problems as aids to thinking and comprehension. *Topics in Learning and Learning Disabilities, 1,* 73–80.

Chapter 5

Curriculum and Instructional Activities Pre-K through Grade 2

Anne Marie Fitzmaurice-Hayes

The scene is a breakfast table. A five-year-old child is holding forth.

"Five and five are ten." She holds up fingers and thumbs of both hands to illustrate.

A pause follows, then a puzzled look. Finally, she says, "Except when you write it down. Then it's a one and a zero." With a shrug, the child turns the discussion to other matters.

This brief episode illustrates the complexity of the concepts with which even the beginning student of mathematics must contend. If a child is to meet this and later challenges successfully, a great deal of groundwork must be laid during the preschool and primary grade years.

In this chapter we will examine the nature of that foundation, and some principles guiding the development of necessary skills and understandings in children with learning disabilities. We will move on to discuss appropriate curriculum topics for the learning-disabled child, instructional strategies, and models for instruction.

INSTRUCTIONAL IMPLICATIONS

Far from being a time of inactivity mathematically speaking, the first seven years of life mark a time of major achievements in learning about quantity.

Tables 5–1 to 5–4 outline the scope and sequence of the mathematics content a child should have encountered, and, for the most part, mastered, by the end of second grade. Any more specific references to age levels have been deliberately omitted since children differ greatly in their pace. The horizontal direction will indicate areas of development that occur at approximately the same time.

Such a scope and sequence serves two purposes. It provides the teacher with a framework within which to plan instruction. The outline can also serve as a basis for assessment.

Table 5–1 Scope and Sequence for Instruction in Whole Numbers, Pre-K through Grade 2

Classification	*Seriation*	*One-One Correspondence*	*Numeration*	*Operations*
Heaping sort				
Graphic sort			Rote counting to 5	
Simple sort	Comparison of continuous quantities		One-one principle for "small" sets	
True sort by perceptual attribute function number	Seriation: length, width, 3 items	Comparison—sets of 5 or fewer members	Cardinality principle for "small" sets	
			Rote counting to 10	
Patterning perceptual attribute category number	First, last	Equivalence—sets of 5 or fewer members	Zero	
	Seriation: length, width, thickness, 4–6 items		Rote counting to 20	Joining of sets and separation of a subset from the larger set
Patterning perceptual attribute category number	Seriation: mass		One-one principle and cardinality principle applied to "larger" sets	
			Expanded rote counting, 100 and beyond	Simple addition and subtraction by counting
				Counting by twos
				Addition: one-digit examples
	Seriation: number	Conservation of number	Place value, tens and ones	Subtraction: one-digit examples
	Ordination	Ordinal numbers and cardinal numbers	Reading, writing numerals	Basic facts
			Place value—100	Two-digit examples
			Even, odd numbers	Subtraction as inverse of addition

Table 5–2 Scope and Sequence for Instruction in Geometry, Pre-K through Grade 2

	Euclidean Geometry	
Topology	*Shapes*	*Relationships*
Open, closed		
Inside, outside, on		
Betweenness	Naming and recognizing two-dimensional shapes: square, circle, triangle, rectangle	
Order constancy		
	Reproducing two-dimensional shapes	Parallelism
	Naming and recognizing three-dimensional shapes: sphere, cube, rectangular solid	Similarity Congruence Symmetry
	Lines Line segments Rays Points	

Table 5–3 Scope and Sequence for Instruction in Fractions, Pre-K through Grade 2

Basic Concepts
Discrete parts
Blended parts
Replaceable parts
Interchangeable parts
Congruent parts of a whole
Names for ½ and ¼
Fractions as parts of geometric regions

Table 5–4 Scope and Sequence for Instruction in Measurement, Pre-K through Grade 2

Metric Geometry						
Length	*Area*	*Volume*	*Angles*	*Mass*	*Time*	*Temperature*
Nonstandard units:						
Height				Comparing masses using a balance		
Length						
Width			"Square corners"			Hot, cold
					Comparing time intervals using nonstandard units	
Distance	Rectangular regions, nonstandard units	Rectangular solids, nonstandard units				
					Days of week	
				Prediction of relative masses	Months of year	Temperature and event
Conservation of length	Nonrectangular regions, nonstandard units	Other solids, nonstandard units			Telling time to the hour and half hour	Seasons of the year and relative temperatures

The treatment of instruction for the child with learning disabilities must be based on a set of principles. The following principles serve to guide the choice of material in the remainder of this chapter:

- Mathematics instruction for young learning-disabled children should systematically embrace many areas of mathematics, those that tap nonverbal abilities as well as those that tap language skills.
- Mathematics instruction for young learning-disabled children should be concept oriented, rather than skill oriented.
- Mathematics instruction for young learning-disabled children should be based on a systematic set of instructor/learner interactions so that each area of strength available to the child can be utilized.
- Care must be taken in the assessment of the young child to ascertain developmental level as well as achievement level.

The instructor of young learning-disabled children will find it necessary to keep the above principles in mind when evaluating mathematics programs.

Instructional Programs

In recent years, several programs embodying a developmental approach to mathematics instruction for young children have found wider use. Space permits us to mention only a few of these. Each of the following programs adheres to most, if not all, of the principles we have discussed.

• Montessori materials (Montessori, 1964, 1965, 1965a) are designed to be hands-on, self-correcting aids to the development of basic mathematics concepts, including seriation, one-to-one correspondence, shape and size, volume and length, counting, place value, addition, subtraction, and more advanced arithmetic skills (Bartel, 1975). The program is self paced, and demonstration of mastery at one task is necessary before a more sophisticated activity is tackled.

• *Early Childhood Curriculum* (Lavatelli, 1973) is a Piaget based curriculum organized around four main themes: classification, number, space and measurement, and seriation. The lessons can be conducted with small groups or individuals. The materials are attractive and sturdy, and many represent objects found in the child's home environment: eating utensils, tools, wooden animals, wooden vehicles, and the like. The approach is a problem-solving one: "The idea in each case is to have the children solve a problem not by judging in terms of how something looks to them, but by doing something physically and mentally to the data, shuffling the facts about in their minds, so that they come to a logical solution" (Lavatelli, 1973, p. 1). A detailed teacher's guide accompanies the program. We should note that, while useful for developing prenumber concepts, the *Early Childhood Curriculum* does little in the way of formal arithmetic.

• *Project MATH, Level I* (Cawley, Goodstein, Fitzmaurice, Lepore, Sedlack, & Althaus, 1976) is a comprehensive mathematics curriculum designed for the preschool or primary grade child with handicaps to learning and achievement. The content covered is divided into the areas of sets (classification), patterns, geometry, measurement, numbers, fractions, and problem solving. Instructional activities are built according to a model called the Interactive Unit, a group of 16 different teacher/learner interactions derived from four teacher behaviors (manipulate, display, say, write) and the corresponding four learner behaviors. The program provides for both pre- and postassessment. It is also accompanied by a set of manipulatives that find frequent use during the instructional activities. The lessons can be conducted with individuals or with small groups; each child's progress is recorded on an individual progress record. The problem-solving component consists of sets of picture materials and story scripts, read by the teacher, that call for the participation of the children in solving simple word problems stressing logical thinking rather than computation.

The teacher who is conscious of the need to develop the content listed in the scope and sequence outline of Tables 5–1 through 5–4 will find many ways to incorporate such instruction into daily activities. The following are suggestions about how such teaching can take place.

Classification

For true classification to take place, the decision to use the rule or scheme with which one sorts a set of stimuli must originate from within the sorter. The origin of the sorting rule and the need to be aware of the different attributes an item may possess are the focus of the following activities.

1. Using advertising circulars or old magazines as the source, help the children collect a number of pictures of food, furniture, tools, clothes, animals, and the like. Shuffle the pictures and give some to each child. Each child is to sort his or her pictures so that all the pictures in a group have something in common. After the children have completed the task, encourage them to explain the groups they have made. As a follow-up activity, use the same set of pictures to arrange a bulletin board display in which you have sorted the pictures. Ask the children to tell you how the items in each of your groups are alike. Then ask the children to suggest other ways in which you could have sorted the pictures.
2. Place a collection of attribute pieces in front of you. Select some pieces that have three attributes in common. For example, take out all the large yellow triangles. Ask the children to tell you all the ways in which the pieces you have chosen are alike (all the pieces are large; all the pieces are yellow; all the pieces are triangles). Add a small yellow triangle to the pile. Ask the

children to tell you what has changed about the way the pieces are alike (all the pieces no longer have the same size). Ask the children to select pieces from the collection in front of you to join to the other pieces so that the group would still have only yellow triangles. The children should tell you to add all the small yellow triangles. Point out to the children that when the pieces are alike in only two ways, more pieces fit the condition than when the pieces have to be alike in three ways. Add a red triangle to the pile. Repeat procedure and discussion.

3. Ask the children to think about, or show them pictures of, a family with four members: mother with light hair, father with dark hair, son with light hair, daughter with dark hair. Discuss with the children the different ways in which the family members might be sorted (young and old, male and female, dark hair and light hair, and so on).
4. Select some children from the class who have more than one attribute in common. For example, call by name all the boys wearing sneakers. Ask them to stand in front of the other children. Ask the children to tell in what way the students in the front of the room are alike. Prompt, if necessary. Repeat the activity by selecting groups of children that have more than two attributes in common.
5. When giving directions for different activities during the day, describe the children in terms of at least two attributes: "I want the students in the first row with dark hair to go to the chalkboard." "Will the girls with blue eyes please line up in front of the room?" "John, please give a piece of paper to all boys wearing sneakers."
6. Seat the children in a circle. Ask the first child in the circle to think of something in the room and to name the color of that item. Ask the second child to think of all the things that have the color named and to name the shape of something in the room that has the color named by the first student. Ask the third child to point to or touch something that satisfies the descriptions given by the first and second children. Continue around the circle, asking two children to name different attributes and a third child to name an item that has those attributes.
7. Discuss with the children the different things that can be bought in a grocery store or in a supermarket. After several items have been named, tell the children that you are going to describe items by listing two sets to which each item could belong. They are to think of something that could be bought in a grocery store or in a supermarket that would satisfy your description. For example, "I am thinking of something that is round and can be eaten. What am I thinking of?" Encourage the children to name several items satisfying your description. As the children become proficient in naming items, restrict the categories to which the items could belong. For example, "I am thinking of a round vegetable that is green."

Patterns

Many different patterns become familiar to children at a very young age. For learning-disabled children, explicit instruction directed toward a search for patterns and recognition of the patterns forming a part of their surroundings is a significant component of the curriculum. Because mathematics is the study of systematic patterns of relationships, the inclusion of formal instruction in patterns in the mathematics program seems to be a most suitable strategy.

A pattern is defined as a repeating sequence of elements, for example, black, white, white, black, white, white, black, white, white. To define a pattern, we must show the sequence at least twice. Patterns can be built around elements that will serve to heighten the child's perceptual awarenesses and aid the child's development in language and logical thought. The latter area includes number itself. In the area of perceptual characteristics, patterns can be built around content that is perceived by any one of the five senses, although for practical purposes classroom activities are usually confined to sight, hearing, and touch, along with movement.

Language patterns are presented by way of pictures as well as words, but the dimensions of color, size, shape, orientation, and topological properties give way to the specific and/or generic names of the items pictured. Number patterns, such as 2, 4, 6, 8, . . . are a special kind of pattern. The elements of the pattern themselves do not repeat themselves, but the rule upon which the progression is based repeats itself, and the prediction of the next number of the sequence is dependent on the discovery of that rule, in other words, inductive thinking.

Children can copy patterns or make them longer. In addition to those two tasks, children can make a copy of a pattern by shifting within the dimension to which the elements of the pattern belong, or shifting from dimension to dimension. For example, the child who knows what patterns are about, when given a supply of blue and yellow attribute pieces or chips, will have no trouble making a copy of a pattern that is represented through red and green pieces.

Early work with patterns will help youngsters develop the conviction that rules governing sequences can be found, and that working from that rule base lessens the need for rote memorization in many cases. Working with patterns affords both the child and instructor many advantages: (1) such exercises provide meaningful and cognitively based experiences for students who are not ready to tackle formal mathematics; (2) working with patterns helps to develop the conviction that rules govern much of what is learned in mathematics; and (3) copying, extending, or composing patterns helps the child to solidify classification skills. Some sample pattern activities appear below.

1. If necessary, explain the meaning and use of the words *smooth* and *rough* to the children. Prior to the activity place items having different textures in a

box or a bag. Prepare the container so that a child can insert his or her hand without seeing what is in the container. Begin the activity by asking a child to reach into the container and to remove an item that matches the texture you specify (smooth or rough). When the child can withdraw items as requested, return all items to the container. Dictate patterns using the words *smooth* and *rough*. The child is to build an example of the pattern you name by withdrawing items from the container and arranging the items in the proper sequence. Encourage the child to rely on the sense of touch.

2. Prepare a worksheet or a transparency on which you illustrate patterns built on open figures and closed figures. Direct the children to copy or to extend each pattern.
3. With the children, compose dance steps of movement patterns. If feasible, match the steps to musical patterns.
4. Encourage the children to collect items from their environment in which patterns are evident: cross-sections of certain fruits and vegetables, some forms of sea life, tile or linoleum patterns, wallpaper designs, and so on. Mount each child's contribution in an appropriate place.
5. With the children, list a number of groups of words that rhyme on a chalkboard or some other visible surface. Encourage the children to compose poems in which the words that rhyme are used. The results may not be prize-winning poetry, but all will have a good time.

Seriation

Experiences in seriation, or ordering members of a set according to some attribute, are crucial to the development of a sound understanding of number. A child must not only learn the number names in the proper sequence but also understand that the number names represent quantities of different sizes, and that the sequence of number names reflects an ordered arrangement of the different sizes. This understanding comes about only with time and much experience in the ordering process. Such experiences will lead up to and include activities involving ordination, the use of numbers to denote position in a sequence.

1. Name or point to an object in the classroom. Ask a child to point to an object longer than the one to which you have pointed. Repeat for shorter and having the same length as.
2. Throughout the day, encourage the children to make and state comparisons, for example, "The straw is longer than this pencil"; "This pencil is shorter than the straw"; "This piece of chalk is shorter than my pencil"; "My pencil is longer than this piece of chalk"; "The math book is larger than the spelling book"; "The spelling book is smaller than the math book."

3. Compare the number of pieces of chalk on the chalkledge with the number of erasers, the number of chairs with the number of tables or desks, the number of adults with the number of children, the number of letters in the name of the month with the number of letters in the day of the week, the number of children who buy lunch with the number of children who bring lunch, and so on.
4. Provide the children with a model of an ordered set of items: sticks, pencils, square regions of different sizes, or other materials that may be available. Provide each child with a set of the same materials. Ask the children to make a copy of what you have done.
5. From a set of five or six items, order the first two or three items according to some dimension. Ask a child to continue to order the items according to the same dimension. For example, if a set of paper dolls of different heights is available, arrange the first two or three according to height. Ask the child to complete the arrangement.
6. From paper, prepare a set of rectangular regions, all of which have the same width but different lengths. Ten regions in all would be sufficient. Put the two shortest regions aside. Arrange the others in order according to length but omit some of the regions from the series, for example, omit the fourth region and the seventh region. Give all the remaining regions to the child, including the two regions you previously set aside. Ask the child to complete the series you began.
7. Ask four students of different heights to stand together. Direct the other children to order the four students according to height.
8. Use a set of items that can be ordered in at least two ways, for example, a set of square regions of different sizes made from different grades of sandpaper, very fine to very coarse. An ordering of the square regions according to size should not be the same as the ordering according to degree of roughness. Order the set according to degree of roughness. Ask the children to name the attribute according to which you have ordered the set. Then ask the children to order the set in another way.
9. When the children line up for some reason, ask the first child in line to raise his or her hand; ask the last child in line to raise his or her hand. Direct all the children to turn around. Repeat the request for the first and last children in line to raise hands. Discuss why first became last and last became first.
10. Arrange a set of pictures in a row. Ask the children to point to the first picture in the row, the second picture, and so on. Rearrange the pictures, and repeat the activity.
11. Provide a set of glasses filled to different levels with colored water, no more than four in number. Ask the children to order the glasses. When the task is completed, direct one child to point to the second in the series, starting from the largest, the fourth in the series, and so on. Then rearrange the

glasses. Ask the learners to point to the first in the series, the second, and so on, without moving the glasses.

One-to-One Correspondence

The ability to recognize one-to-one correspondence is fundamental to the understanding of number. For young children, sets with no more than five members should be used. Gradually, the number of items in the sets should increase. Most primary grade mathematics books provide some experiences in one-to-one correspondence. A few suggestions for manipulative activities are listed here.

1. Divide the children into two groups and give each group a set of blocks or any other small items. Instruct the members of one group to display a set of their items. The members of the other group are to construct a set that is equivalent to that displayed by the first group. The first group checks the work of the second group by pairing the members of the two sets to demonstrate the one-to-one correspondence. The two groups then reverse roles.
2. Help the children to make hangings of paper shapes by cutting out the shapes and attaching a number of shapes to a piece of string. Hold up two strings and ask the children if the sets show one-to-one correspondence. A child can check his or her response by drawing a picture of the two sets of shapes and drawing paths between each member of one set and the corresponding member of the other set. The check can also be performed by attaching the two strings to the chalkboard and drawing paths from the members of one set to the members of the other set.
3. Give each child a sheet of paper folded into sections. Assign the children to work in pairs, or allow the children to choose partners. Give the children in each pair a collection of gummed stars. The partners are to work together in this way: one child makes a set of gummed stars in one block on his or her paper; the other learner is to draw a set with as many members in the same block on his or her paper. When the children are satisfied that their sets match, they are to exchange roles and continue this procedure until they have sets in all the blocks on the paper.
4. Help the learners to prepare index cards on which are pasted gummed stars or shapes. The number of shapes or stars should vary from one to five, and there should be four cards for each number from 1 to 5. In addition, one card should be left blank. The children can use the cards to play a variation of "Go Fish." The children can also use the cards when working independently. Tasks might include sorting the cards by number, arranging the cards in sequence so that four sequences are formed, or drawing pictures of

sets equivalent to those represented on each of four cards drawn at random from the deck.

Numeration

Primary level textbooks abound with teaching suggestions in the area of numeration. Suggestions here will be limited to those that might be particularly helpful to the instructor responsible for helping learning-disabled children achieve a better grasp of what numbers are all about.

1. Initial counting experiences should be structured so that the child has a maximum chance of succeeding. Much practice in the recitation of the number names in their proper order should be provided before these number names are employed in the actual counting process.
2. To begin with, the child should count movable items arranged in a row. This setup allows the child to physically employ the one-one principle and the cardinality principle referred to earlier in this chapter. Gradually, other types of sets of things can be introduced:

   - Groups arranged in rectangular arrays.
   - Pictures of things arranged in a row or a column. Note that for these groups the child cannot physically move each item as he or she counts it. Other strategies have to be used. For example, a child could use a crayon to mark each item in some way as it is counted.
   - Pictures of arrays.
   - Groups of objects randomly arranged. In this case, the counter is almost forced to move an item as it is counted, since no pattern exists in the arrangement.
   - Pictures of groups of objects randomly arranged.
   - Repeated actions, done by others (as one child hops, another child counts the number of hops) or done by self (child hops and counts).
   - Repeated events. For example, the number of animals sighted during a walk through a park.
   - Repeated noises. For example, the chimes of a church bell, fire whistles, and so on.

3. Build sets having a designated cardinal property and ask the children to name other sets that have the same number of elements. For example, make a set composed of two animals and a tree (using pictures or models). Ask the children to name the number of items in the set and then to name the members of another set that would have the same cardinal property.

4. Build two sets that differ in their members or in their cardinal property or both. For example, one set might be made up of pictures of a car, a truck, and a van. The second set might show pictures of an apple, a grape, and an orange. Ask the children to tell you how the two sets are alike and how they are different.
5. Name a number between 1 and 10. Direct a child to make as many sets as he or she can to represent the given number.
6. Display several sets so that each of four numbers between 1 and 10 is represented at least twice. Name a number between 1 and 10. The child is to point to all the sets representing that number. If no such set exists in the display, the child should not point to any set.
7. Initial instruction in place value can take place while counting loose items. Children should be encouraged to group items in sets of ten as they count. This procedure offers two advantages. It provides a strategy for organizing the items that have been counted so that if the counter is distracted during the counting activity the task need not be begun from the very beginning. The practice also helps the child to become conscious of groups of ten very early in her or his experiences with numbers.
8. For learning-disabled children, the use of manipulatives as aids to understanding place value should prove to be a valid and valuable technique, although the nature of the manipulatives bears scrutiny. Some manipulatives offer a closer resemblance to the meaning of place value than others. The use of popsicle sticks or tongue depressors, or similar materials (toothpicks, pipe cleaners, straws, etc.) is advantageous in the first stages of place value instruction. The materials can be grouped in sets of ten and the sets bound by rubber bands or the ties used for closing plastic bags. In this way one or more sets of ten can be represented, and the link to the symbols can be made. The equality between ten ones and one group of ten can be illustrated over and over again by grouping the single units or breaking apart an already established set of ten. After much practice in representing two-digit numbers, the children can write matching two-digit numbers. The use of these same items can afford the children a means of representing the processes of regrouping and renaming in the context of addition and subtraction. We will discuss other manipulative aids for developing the concept of place value in the next chapter.

Operations

Early experiences in addition and subtraction should consist of repeated practice in joining sets and separating a subset from a set. In each case counting should be used to note the size of each set involved, and comparisons should be made between initial sets and final sets. Children should be encouraged to use whatever strategy they find most efficient to keep track of what is going on. Above all, children should be convinced of the value of using counting as a strategy until repeated experience has provided them with ease in memorizing the different combinations. Guessing sums and differences should be avoided. Every time a child guesses and states or writes an incorrect sum or difference, he or she rehearses an incorrect behavior. Gaining facility in the operations of arithmetic is one area in which haste definitely does make waste. Learning-disabled children have very little academic time that can be wasted.

Geometry

Geometry offers a wealth of opportunities for providing children with practical, everyday experiences with mathematics and at the same time developing skill in inductive thinking. Children who are not yet ready for formal arithmetic instruction will often profit from several good units in geometry. The following activities are but a few of the many possibilities that exist.

1. Show the children a display of three or four pictures, all of which, except one, show a closed container (for example, closed jars and boxes). The remaining picture should show an open container. Ask the children to find the picture of the container that is different from the others. Repeat the activity for other topological attributes, inside, outside, on, and order constancy.
2. Use masking tape, heavy yarn, or rope to outline two closed areas on the floor or the surface of a table. Inside one area place a number of red items (attribute pieces, pieces of colored paper, or the like). On the boundary of that same area place a number of yellow items. Outside the area (but not inside the other area) position a number of blue items. Give each child a set of items of different colors. Direct the children to make the second area look like yours. When the direction has been carried out, discuss with the children the attributes the items inside the areas have in common: they are red; they are inside. Repeat for outside and on.
3. If the group is large, seat the children in a circle; otherwise a semicircle or straight line arrangement will do. Begin the activity by posing riddles such as the following: "I am thinking about the child sitting between Sean and Irene. Who am I thinking about?" After a few such riddles, expand the descriptions to include items in the classroom that can be described in terms of betweenness.
4. Place four or five beads of different colors on a string. Ask the children to name the colors of the beads in the order in which they appear on the string. Change the position of the string but do not restring the beads. Again, ask the children to name the colors of the beads in order. Repeat for several positions

of the string. Then ask the children to predict what the order of the beads will be after you move the string again.

5. Work with the children to make a small model of a train with cars of different colors, using whatever materials might be available. Have on hand a paper towel tube or a similar tube to represent a tunnel. Discuss the order in which the cars of the train are lined up. As the children watch, start the train and move it into the tunnel. Stop the train before it begins to emerge from the tunnel. Ask the children to predict the order in which the cars of the train will emerge from the tunnel.
6. Before the lesson, place representations of squares, circles, triangles, and rectangles in different locations in the classroom. To conduct the lesson, show the children a model representing one shape. Ask them to find the models of the shape like the shape demonstrated. The same exercise can be carried out for three-dimensional figures.
7. Direct the children's attention to a standard set of models for upper-case and lower-case printed letters. Near the display of letters place representations of parallel lines. Ask the children to pick out examples of parallel line segments in the letter formations. Repeat for pairs of perpendicular line segments.

Fractions

Young children become aware of fractions in a limited, intuitive way. Formal work with fractions should not make up a large part of the young child's instruction in mathematics, but certain understandings about the part-whole relationship can pave the way for greater facility in work with fractions. To that end the following activities are suggested:

1. Using a puzzle of the human body, parts of an animal, parts of a flower, or the like, discuss with the children the relationship of each part to the whole. Discuss also the fact that each part is a whole in and of itself, and that parts may have parts. Thus, the hand of the human body has as its parts the fingers, thumb, and palm.
2. Demonstrate and discuss with the children objects that have interchangeable or replaceable parts: cars and tires, a bicycle wheel and spokes, shoes and laces, and so on. Note that replacement parts usually must be of the same size and shape as what is being replaced.
3. Work with the children to prepare a set of fraction puzzles in the following manner. Using cardboard from carton boxes, make circular regions with diameters no less than eight inches. The number of regions you need will depend on the abilities of the children with whom you are working. Divide one region into halves, one region into thirds, one region into fourths, one region into sixths, and so on. Cut only one fractional part from each region.

Place the circular regions with one part missing on a table, and randomly arrange the missing pieces on the same table. The task is to complete each circular region with the correct piece. Observe each child's strategy in performing the task.

Measurement

Measurement is one of the branches of mathematics that is most applicable to daily living, yet it receives far less attention than it deserves. The young child becomes aware of certain measures early in his or her experience: weight, height, time, temperature, distance. The child's understanding of the quantification system we bring to bear on those dimensions of our existence is very primitive, but it contains the seeds on which later concepts will be built. The comprehension of measurement seems to be very closely linked to development; for this reason, early measurement activities must be monitored carefully, so that the child is not forced to deal with abstractions beyond his or her capabilities.

The activities listed are intended to give a taste of what measurement activities for the kindergarten, first grade, or second grade child should encompass. The mathematics programs discussed earlier in this chapter provide many measurement activities for the young child. In addition to those sources, *Young Children Learning Mathematics* (Cruickshank et al., 1980) will be helpful.

1. Gather a set of pictures illustrating events that take place during the day or during the night. Arrange the pictures in random order on the surface of a table. Ask the children to sort the pictures into two groups, one group consisting of events that take place during the day and the second group consisting of events that take place during the night. At another time, ask the learners to sort the pictures into three groups, one for morning events, one for afternoon events, and the third for night events. At this time you may wish to introduce the terms midnight, noon, a.m., and p.m.
2. Cover the clock face if it is within view of the children. Use a watch with a second hand or a stopwatch for the activity. Ask the children to perform different actions for stipulated amounts of time. For example, tell the children that they will be clapping their hands for one minute, walking around the room for two minutes, reading or looking through a book for five minutes. Discuss with the children their feelings about the task: did the time go by slowly or quickly? During another lesson, repeat the activity but ask the children to be the judges as to when the time for the task has expired. Again, discuss the results and the children's feelings.
3. Collect a set of pictures showing different events taking place outdoors (swimming, skiing, a barbecue, sledding, leaves falling from a tree, and the like) and pictures illustrating different kinds of clothing. Ask the children to

sort the pictures into groups that show events taking place and clothing worn at different times of the year. At another time ask the children to match each event with the type of clothing that would be worn for that event. When the names of the months of the year and the seasons of the year are being learned, the children can match each picture with the name of the month for which it is appropriate, or the name of the season of the year during which that event would most likely take place.

4. Show the children a ruler one foot in length and a yardstick. Ask the children to find objects in the classroom that would be about as long as the ruler or the yardstick. Place the ruler and yardstick where the children can see them, but encourage the children not to physically compare the objects they have collected with the standards until the activity is completed. When each child has selected some items, allow them to use the ruler and yardstick. Discuss the reasons for any large discrepancies.
5. Point out a movable object in front of the classroom and one in the back of the classroom to the children. The objects should be such that they are not of the same height, but the difference should not be too great. Ask the children which of the objects is the taller. If the objects are close enough in size, a discussion should ensue. Ask the children how the question might be resolved. Answers might include moving the two objects so that they are adjacent to each other. If that seems to be the agreed-upon solution, select two other objects, again similar in height, but this time the objects should be such that they cannot be moved. Repeat the activity, again forcing the question of how the question might be answered accurately and with assurance. Lead the discussion to the need for a movable third item that can be used for measuring each of the nonmovable objects.
6. Body maps offer a good opportunity for children to observe their own height. Obtain a quantity of large paper, enough so that each child can lie on a sheet of paper. Assign each child a partner. Each child is to trace an outline of his or her partner's body while the partner is lying on a sheet of paper. The children can then color in the different parts of their bodies and clothing and cut the life-size figure. These can be used for many comparison activities.
7. Pose questions to the children that encourage them to think about relative lengths, heights, volumes, areas. Examples: "Could an elephant fit through the classroom door?" "Could the desk fit into the closet?" "Could an apple pie fit into a lunchbox?" "Could this pile of paper clips fit into that box?" "Will this book fit into that drawer?"

Problem Solving

The developmental approach to the teaching of mathematics to young learning-disabled children lends itself easily to an emphasis on the importance of problem-solving activities. This focus demands that, whenever possible, new material is to be approached through the posing of dilemmas that require the child's active participation for solution. Such an activity was the fifth one in the measurement section. Most parents will testify to the ingenuity of young children when they are faced with a problem they wish to solve. If children are poor problem solvers when they grow older, somewhere along the line this creativity has become stifled.

Problems can be either verbal or nonverbal in nature. That is not to say that nonverbal problems have no language requirements, but rather that the language component is less formally necessary to the solution. Both types of problems should be included in the experiences of young learning-disabled children.

The activities below represent a sample of both types of problem situations. *Project MATH* (Cawley et al., 1976) offers an excellent program in word problems for young children. As noted earlier, *Early Childhood Curriculum* (Lavatelli, 1973) is based on a problem-solving approach to the development of early mathematics concepts. The teacher who is aware of the need for young children to build on their natural abilities as problem solvers will find many opportunities during the school day to encourage this development.

1. On sheets of 9″ × 12″ drawing paper, work with the children to prepare pictures that will form the bases for stories. For example, one picture could represent the local fire station, another the town library, a third the post office. Ask the children to collect small pictures of items that might go with each of the larger scenes: fire engines, hoses, ladders, books, envelopes, stamps, packages, and so on. Another set of large pictures could consist of vegetable gardens, with small pictures of different foods that can be grown in a vegetable garden. These materials can be used in the context of problem stories that the teacher reads while the children use the small pictures and the large pictures to act out the solutions to the problems posed. Such problems need not be merely quantitative in nature. Questions can be posed in such a way that the children are forced to use elementary logic to obtain the answer. For example, assuming the vegetable gardens form the substance of the story:

   - Members of the cooperative planted three pepper plants, four tomato plants, and two bean plants. How many of the plants were not pepper plants?
   - Members of the cooperative planted four potato plants, three squash plants, and four marigolds. How many vegetable plants did the members of the cooperative plant?
   - In one vegetable garden, the members of the cooperative planted two pepper plants, five heads of lettuce, and two nasturtiums. In another garden, the members of the cooperative planted three broccoli plants, five

marigold plants, and four melon plants. How many vegetable plants did the members of the cooperative plant in all? How many flower plants did the members of the cooperative plant in all? How many plants did they plant that were neither flowers nor fruit? Did the members of the cooperative plant more vegetable plants or more fruit plants?

2. Select a flat surface, such as the top of a table or an area of the floor bounded by masking tape outlining a rectangle. Give one group of children a set of small tiles or blocks, all of the same size. Give another group of children another set of tiles or blocks, also all of the same size, but larger than the tiles or blocks with which the first group will be working. Each group of children, in turn, is to use the tiles or blocks to find out how many such pieces it would take to completely cover the rectangular region to be measured. When each group has completed the task, discuss the results, posing questions like these:

    - Why did the results differ?
    - If a third group of students were to do the same task with yet a third set of tiles or blocks, would their results be the same as either of the first two groups? Why or why not?
    - If the tiles used by the third group were smaller than those used by either of the first two groups, would the number of tiles required to cover the rectangular region be larger or smaller than the number of tiles used by either of the first two groups?

3. On the floor or table, outline two circular regions so that they overlap slightly. In one region place a number of items, all of one color, for example, blue. In the other region place a number of items, all of which are of different colors, but all of which are circular in shape. Put nothing in the area where the two regions overlap. Ask the children to name or point to items that would appropriately be contained in the overlap.

REFERENCES

Bartel, N.S. (1975). Problems in arithmetic achievement. In D. Hammill & N. Bartel (Eds.), *Teaching children with learning and behavior problems*. Boston, Allyn & Bacon.

Brigance, A. (1977). *Brigance diagnostic inventory of basic skills*. Woburn, Mass.: Curriculum Associates.

Cawley, J.F., Cawley, L., Cherkes, M., & Fitzmaurice, A.M. (1980). *Beginning Education Assessment*. Glenview, Ill.: Scott, Foresman.

Cawley, J.F., Goodstein, H.A., Fitzmaurice, A.M., Lepore, A.O., Sedlack, R., & Althaus, V. (1976). *Project MATH. Level I*. Tulsa, Okla.: Education Development Corp.

Lavatelli, C.S. (1973). *Early childhood curriculum*. Cambridge, Mass.: American Science and Engineering.

Montessori, M. (1964). *The Montessori method*. New York: Schocken Books.

Montessori, M. (1965). *Dr. Montessori's own handbooks*. New York: Schocken Books.

Montessori, M. (1965). *The Montessori elementary materials*. Cambridge, Mass.: Robert Bentley.

Chapter 6

Curriculum and Instructional Activities Grade 2 through Grade 4

Anne Marie Fitzmaurice-Hayes

Children's experiences with mathematics can be and should be both exciting and rewarding. Learning-disabled children should be no exception to that maxim. Grades 2 through 4 can serve as the setting for the building of well-established skills, good concepts, and positive attitudes toward mathematics.

In this chapter we will list appropriate curriculum topics for the learning-disabled child in grades 2 through 4, and discuss instructional programs and strategies.

INSTRUCTIONAL IMPLICATIONS

The child of seven or eight years of age has also acquired certain specific concepts and skills in mathematics. If previous mathematics instruction has been appropriate, most children in grade 2 can do the following:

1. count rotely to 100,
2. count up to 24 items,
3. join sets,
4. write numerals to 100,
5. read number words to *five*,
6. use several ordinal numbers,
7. read numerals to 999,
8. write numerals to 999,
9. count by twos to 20,
10. recognize and understand a few fraction names,
11. memorize basic addition and subtraction combinations,
12. add and subtract two and three digit numbers without renaming,

13. add and subtract two digit numbers when only one renaming is necessary,
14. recognize and state value of coins,
15. tell time to the quarter hour,
16. recite days of the week and seasons of the year,
17. recognize a thermometer,
18. recognize and reproduce familiar geometric shapes, and
19. recognize familiar three dimensional shapes. (Brigance, 1977)

In addition, these children can solve word problems calling for categorization, union, separation, exclusion of extraneous information, and negation (Cawley, Goodstein, Fitzmaurice, Lepore, Sedlak & Althaus, 1976). The list represents no mean accomplishment. Instruction in grades 3 and 4 can build on all of these concepts and skills.

Tables 6–1 through 6–4 outline the scope and sequence of the mathematics content a child should learn by the end of the fourth grade. The scope represents a middle-of-the-road approach; with good instruction learning-disabled children can learn most, if not all, of the content listed. For the child who seems to have a serious disability in one or more areas of mathematics, care must be taken that successful performance takes place in areas not marked by disability. Thus, the child who has great difficulty in reading or writing numerals may be able to do quite well in other areas if the need for written numerals is controlled.

The remainder of this section is devoted to specific concerns and instructional strategies in the different areas of mathematics. Our selection of topics is based on two sources. Some represent questions or concerns most frequently voiced by teachers with whom we have worked over the years; others reflect material that is significant to a child's overall progress in mathematics.

Whole Numbers, Ordinal Numbers

We use numbers not only to describe a given quantity, but also to designate position in a sequence. Numbers used in that context are called ordinal numbers. The situations in which ordinal numbers are used give us some idea of which of them need to be mastered. In addition to using ordinal numbers to designate the sequence of days in a month, we need the first ten ordinal number names most often. The child should master both the numeral (1st, 2nd, . . .) spelling and the letter spelling of these names.

Other concepts and skills also require attention. The first of these is the understanding of orientation or direction. The first child in a line of students facing in one direction is the last one in line if everyone faces in the opposite direction.

Table 6-1 Scope and Sequence for Instruction in Whole Numbers, Grades 2–4 (7–9 Years)

Classification	*Seriation*	*Numeration*	*Operations*
		Conservation of number	Inverse relationship, addition and subtraction
	Ordination	Place value, 2- and 3-digit numbers	
		Number line	Addition and subtraction, up to 3-digit numbers, one renaming
		Place value, 4- and 5-digit numbers	Commutative and associative principles for addition
	Ordinal number names through tenth, numerals (1st, 2nd, etc.)	Inequalities	Basic facts, beginning memorization
		Place value, hundreds, thousands, millions	Addition and subtraction up to 4 digits, two renamings
Hierarchies of classification	Ordinal number names through tenth, words	Reading, writing numerals having 2 to 7 digits.	Multiplication, basic concepts and facts
		Counting by 3s, 4s, 5s, 6s, 7s, 8s, 9s	Commutative and associative properties of multiplication
		Rounding off numbers	Distributive property of multiplication over addition
		Multiples	
		Factors	Multiplication, 2- and 3-digit factors, no renaming
		Prime factors	
			Multiplication, one renaming
			Division, basic concept

Table 6–2 Scope and Sequence for Instruction in Fractions, Grades 2–4 (7–9 Years)

Numeration	*Operations*
Fractions as names for parts of geometric regions	
Fractions as names for parts of sets	
Equivalent fractions	
Mixed numerals, improper fractions	
Different names for the same fraction	
	Addition, subtraction of fractions with like denominators
Use of decimal fraction to write fractions with denominator of 10	
Fractions and linear measure	

Table 6–3 Scope and Sequence for Instruction in Geometry, Grades 2–4 (7–9 Years)

Euclidean Geometry		
Shapes	*Relationships*	*Coordinate Geometry*
Points		
Lines		
Line segments		
Rays		
Angles		
Quadrilaterals		
Square		
Rectangle	Perpendicularity, lines	
Parallelogram	Parallelism, lines	
Rhombus		
Triangles		
Right triangles	Congruence	
Circles		
Diameter	Symmetry	
Radius		
	Similarity	
Rectangular prisms		
Triangular prisms	Rigid motions, translations, rotations, reflections	
Cones		
Pyramids	Perpendicular planes	
Edges, faces, vertices of 3-dimensional figures		
	Parallel planes	
		Locating points in the coordinate plane

Table 6–4 Scope and Sequence for Instruction in Measurement, Grades 2–4

	Metric Geometry							
Length	*Area*	*Volume*	*Angles*	*Mass*	*Time*	*Temperature*	*Money*	*Liquid*
Conservation of length							Identifying coins to half dollar	
Foot, inch, yard								
Meter, centimeter				Using a balance scale	Telling time to the minute	Reading thermometers, Celsius and Fahrenheit		
Applications of transitivity					Relationships among time units		Relationships among coins	
	Counting square units						Counting change to one dollar	Conservation of liquid
			Recognizing right angles, acute angles, obtuse angles	Conservation of substance				
Mile		Counting cubic units			Using calendar to determine day of week			
Kilometer				Conservation of mass			Buying, selling, saving, spending	Recognizing liquid measures, cup, pint, quart, gallon, liter, milliliter
Measuring length to ½″ and ¼″				Pound Kilogram				
Perimeter								Relationships among liquid measures

This concept can be developed in very simple ways, formally and informally. Suggestions:

1. Arrange some pictures in a row. Ask the children to point to the first picture in the row after you have established a direction. Ask the children to point to the second picture, the third picture, and the last picture. Rearrange the pictures, name the new direction, and repeat the task.
2. Movement can be a good source of activities for increasing familiarity with the importance of orientation for ordinality. Marching games, with the use of "about face," can serve to make children aware of the need to keep direction in mind when orders are to be followed.

Using ordinal numbers successfully requires that the user realize the relationship between the cardinal numbers and ordinal numbers, and that the user be able to translate a number from one type to the corresponding number from the other type. Thus, the child must know that both *one* and *first* represent the starting point for a sequence of numbers, and that *one* is translated to *first* when the context is one of ordination. The following activities should facilitate both the matching aspect of using ordinal numbers and the translation process.

1. Help the children to prepare several pictures of fish by cutting out the shapes from construction paper. Attach to each fish a cardinal number from 1 to 10, one number to a fish. Prepare a chart by listing the ordinal number names in order from top to bottom. Give each child a "fishing pole." When a child catches a fish, he or she matches the numbered fish with the correct ordinal number name on the chart.
2. Help the children to cut out frames of comic strips and to paste them on oaktag or some other heavy piece of paper or cardboard. Distribute a different comic strip to each child. He or she is to arrange the frames in the correct order and label each frame with the correct ordinal number name for its position in the sequence.
3. Using masking tape or chalk, draw a representation of a number line on the floor and divide the number line into ten segments. Do not number the segments, except to write 0 at the beginning of the number line. Give cards on which are printed the cardinal numbers to each child in one group and cards on which are printed the ordinal number names to each child in a second group. Ask a child from the "cardinals" to take his or her place on the line; then ask a child with the corresponding ordinal number name to take his or her place. Repeat until all the cards are matched.

Whole Numbers, Place Value

Our system of numeration is an efficient one, but that very efficiency poses problems for children attempting to learn how to cope with larger numbers. The decimal system uses groups of ten to help us keep track of how many *many* is. The position of a digit in a numeral names the number of groups of ten with which we are dealing in any given circumstance. Children must understand both those features if they are to be successful in working with numbers.

Initial instruction in place value involves counting loose items and use of manipulatives, techniques described in Chapter 5. As stated, popsicle sticks, tongue depressors, or small coffee stirrers are effective in the first stages of place value instruction, since the materials can be grouped in sets of ten and the sets bound by rubber bands or the ties used for closing plastic bags. In this way one set of ten or more can be represented, and the link to the symbols can be made.

A step removed from the popsicle sticks are materials such as chips of different colors, one color for each power of ten to be represented. Thus, blue chips might represent ones, red chips might represent tens, yellow chips hundreds, and so on. The advantage of these materials is readily apparent. Imagine representing 386 using popsicle sticks. On the other hand, chips do not visually represent numbers in the same way that popsicle sticks do. That is to say, for the number 386, the child using chips will see only 17 chips altogether, 3 yellow ones, 8 red ones, and 6 blue ones. If chips are to form the basis for manipulative representations of numbers, the children must have an opportunity to practice using them, and the rules of representation must be made very clear. Many experiences in representing two-digit numbers and three-digit numbers should be provided. Also necessary is practice in exchanging chips: a ten chip for ten ones chips, ten ones for one ten, ten tens for one hundred, one hundred for ten tens, and so on. Practice at this point of instruction will help with respect to regrouping and renaming later on.

Several commercial programs using manipulatives designed especially to aid students in the development of the concept of place value are available: Cuisenaire rods, Dienes Multibase Arithmetic Blocks, and the Stern Structural Arithmetic Apparatus. *Project MATH* (Cawley et al., 1976) includes both chips and attribute pieces; the Montessori materials contain several components to aid the understanding of place value.

A time-honored tool for both the representation of numbers and calculations with them is the abacus. As shown in Figure 6–1 the abacus is constructed in such a way that the counters used to represent different powers of ten are built into a frame. The problems associated with loose chips or blocks can be avoided. If necessary the instructor can label each column with the name of the power of ten represented by that column. Later on in the instructional sequence, a decimal point can be inserted between any two columns and the abacus can be used to represent operations with decimals and whole numbers.

Figure 6–1 A Typical Abacus

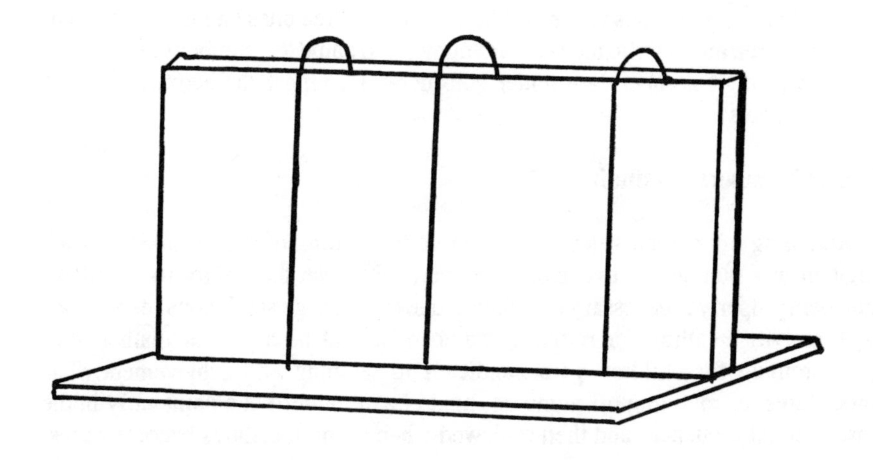

The following exercises represent but a sample of the activities that can and should be conducted over a period of time to ensure that children attain a mastery of place value.

1 Write a two-digit numeral on the chalkboard. Provide the children with popsicle sticks or coffee stirrers. Ask the children to represent the numeral you have written. Repeat the activity for several two-digit numerals. When three-digit numerals are encountered, repeat the activity for those numerals.
2. Using popsicle sticks or coffee stirrers, represent a two-digit numeral. Ask the children to write the numeral you have represented. Then ask the children to state the name of the number you have represented. When three-digit numerals are encountered, repeat the activity for those numerals.
3. If the children are thoroughly familiar with the value of a dime and the value of a penny, and therefore the relationship between the two, represent two-digit numbers using dimes and pennies. Ask the children to state the name of the number you have represented. Ask the children to write the name of the number you have represented.
4. Write an expression on the chalkboard such as 3 tens + 14 ones. Provide the children with popsicle sticks or coffee stirrers. Ask the children to represent the quantity you have represented in two ways. The children should show 3 tens and 14 ones, and 4 tens and 4 ones.
5. Write a numeral on the chalkboard such as 43. Ask the children to represent the same quantity in as many ways as possible, using chips that represent

ones and chips that represent tens. Responses might include representations of 4 tens and 3 ones, 3 tens and 13 ones, 2 tens and 23 ones, and so on.

6. Dictate expressions to the children such as "three ones and five tens." The children are to write the standard numeral to name the number given. When three-digit numbers are under consideration, repeat the activity for such numbers.

Whole Numbers, Rounding Off

Rounding off numbers depends on an understanding of the meaning of each digit in any number of two places or more. Because the ability to round off numbers properly is necessary to estimate answers during calculations, experience with activities calling for rounding numbers should be a part of mathematics programming for children with handicaps to learning and achievement. The procedures for rounding off numbers should be introduced at a point early in the instructional sequence, and then reviewed when such procedures become necessary for estimation.

Skill in rounding off numbers finds many applications. Distances from one place to another are seldom expressed in terms of fractions of a mile but rather in terms of the nearest whole number. Distance recorded by the odometer on a car is rounded off to the nearest tenth of a mile. One's weight is usually expressed in terms of the nearest whole pound. Even the IRS is willing to accept figures rounded off to the nearest dollar. Skill in rounding off numbers is essential if one is to interpret prices like $9.99, $89.95, $399.99 properly.

There is one other strong argument for teaching learning-disabled children, from the primary grades onward, how to round off numbers. The use of a calculator for mathematical computations need no longer be justified. The calculator and desk-top computer are here to stay. As educators our role is to prepare children for the best possible use of these instruments. That is not to say that there is no longer any need for children to memorize the basic addition and multiplication combinations, or to learn the basic computational processes. To use a calculator efficiently, one must be able to perform the mechanical steps necessary for a computation, and then read one's output critically to see if it in fact represents a probable accurate outcome. That judgment requires both a store of answers to the basic combinations in multiplication and addition, and a knowledge of how to estimate.

The skill of rounding off numbers must be approached slowly, and much practice must be done at every step in the sequence. Locating numbers on a number line and gauging and comparing the distance from the location of a two-digit number to each of the nearest multiples of ten to determine which of the distances is the shorter one is an effective method of presenting the concept of the "nearest ten." The same technique can be used to illustrate the meaning of the "nearest hundred." After much practice in locating numbers on a number line to determine the nearest ten and/or the nearest hundred, learners can be encouraged to develop the rules for rounding off for themselves.

Exercises in this area can be presented through the use of miniproblems. Questions of the following types are suitable for such practice: About how many children go to your school? About how many people live in our city or town? About how far is it to Chicago? Another source of practice in rounding off numbers is a page from an advertisement circular showing prices on which the children can use their new-found consumer skills.

Whole Numbers, Addition

As a preface to the consideration of the operations of addition, subtraction, and multiplication, a distinction between the mathematical definitions of the operations and our customary ways of thinking about them must be mentioned. Each of the operations on whole numbers has a formal meaning, if not its own, then one related to another operation. Thus, we have learned that subtraction is the inverse of addition, and division is the inverse of multiplication. These formal definitions are good and necessary, but for the learning-disabled child, understanding of the operation as it is used practically must evolve before the more formal definitions appear on the scene. This is the approach that will dominate our discussion of the operations on whole numbers.

Rules for computation are called algorithms. Most mathematics programs teach the standard algorithms, those most of us learned and use today. There are, however, alternate algorithms for each of the operations. These rules for computations can sometimes be of more value in the instruction of learning-disabled children. For that reason, along with activities designed to help children understand the meaning of the operations, we will discuss alternate algorithms.

The understanding of addition is perhaps the easiest among the four operations. Briefly, addition is a symbolic way to tell what happens when we join two sets that do not have any members in common. Addition is an operation we perform on numbers. The process provides us with a shorthand way in which to tell the number story of what happens when we join sets. We do not add sets, or things. We add only numbers.

Most, if not all, commercially produced mathematics programs include a scope and sequence chart. Many times this chart forms part of the Teacher's Guide. Many publishers also prepare scope and sequence information as part of their advertising programs. For those special educators responsible for part or all of a child's mathematics program, familiarity with the texts, including an awareness of scope and sequence, used in the regular school program will be beneficial. This information may not only serve as a guide to the student's instruction but also provide another link between the special education personnel and the regular class teacher. This link offers good potential for improving both the quality of the youngster's program in the special education classroom and the transition back to the regular program when such a return becomes desirable.

An introduction to addition should be effected through exercises in joining sets; in those exercises the children should note the number of items in each individual set and the number of items in all when the sets are joined. The larger number, then, can be defined as the sum of the cardinal properties of the two smaller sets. Much of this experience should be available in first and second grades. Through repetition of such activities, groundwork is laid for the memorization of the addition facts, which should take place gradually, and in the context of meaning.

The addition combinations can be represented on a number line also. The format of such an activity can correspond to the ages of the children. For young children a number line drawn on the floor can be adapted to many types of games. For older students a tape measure or a part of one can be taped horizontally or vertically on a chalkboard, bulletin board, or even across the top of a table or desk. When using the number line, children should start at the point representing the first addend under consideration and move to the right the number of units named by the second addend to arrive at the sum in a two-addend example.

The child who can use manipulatives to represent addition sentences and who can name an addition combination when it is represented is ready to memorize the combinations. With proper instruction most learning-disabled children can commit the basic facts to memory. Good instruction requires that the child be able to experience many forms of drill and practice on the combinations. Different forms of practice are listed below:

1. Call out pairs of single-digit numbers from addition combinations on which the children are working. The children are to call out the sums loudly and clearly.
2. Seat the children in a circle. The first child to your right is to call out a single-digit number from a list provided. The child next to him or her is to call out another number from the same list. The third child is to name the sum of the first two numbers called. Continue around the circle in the same fashion.
3. Obtain an oven timer, an egg timer, or some other timing device. Determine with the child(ren) the number of combinations that are to be dealt with in a specific number of minutes. For example, you and the child(ren) may decide to work with seven addition facts and to allot four minutes to the task. Set the timer. State two addends. The child(ren) must state the sum. Continue until all seven facts have been covered, or until the timer sounds the completion of the time allowed. Repeat the procedure several days in a row, until the time needed for all seven combinations meets the expectations of you and the child(ren).

4. Vary the activity described above by using the timer to determine how long the child(ren) take(s) to answer all seven examples, in writing or orally. Record that length of time on a chart. Repeat each day, or several times a day. Using a line graph or some other representation, record the progress in naming the combinations.
5. State the combinations in the form of riddles such as "I am a number that is three more than five. What number am I?" Encourage the children to make up their own riddles.
6. Many teachers use a board game, such as Bingo, to provide practice in the addition combinations. In such games, care must be taken to minimize the effects of "losing."
7. Help the children to prepare a deck of cards. Select ten addition facts on which practice is needed. Write each expression, without the sum, on a card, one card per expression. Write the computed form of each expression on another ten cards. Write the number representing the sum of each expression on two cards. Thus the deck will have forty cards, ten sets of four like the following: $3 + 2$, $2 + 3$, 5, 5. Two, three, or four children can use the deck of cards to play a variation of "Go Fish."
8. Provide each of several children with cards on which are written sums for different pairs of single-digit addends. Call out an addition combination. The children with cards on which the sum is written hold up the cards. A referee, the teacher or another child, judges the accuracy of the responses and awards two points to each learner holding up an appropriate card. Continue the game until a child is awarded a predetermined number of points.
9. Prepare a set of flashcards on which addition facts needing drill and practice are written. On one side of the card write the complete number sentence. On the other side of the card write the incomplete number sentence. Show the child each complete number sentence and ask the child to read the entire sentence aloud. Repeat this procedure a few times. Then show each complete number sentence again, but this time require only that the child name the sum aloud. Show each sentence again, but cover the sum with a small piece of cardboard. Ask the child to name the sum aloud. If the child hesitates, uncover the sum, ask the child to read the whole sentence aloud, ask the child to read just the sum aloud, and then cover the sum again and ask the learner to name it. Repeat this procedure until the child can successfully complete each number sentence. Then show the child the incomplete number sentences, asking the child to name the sums aloud. After this point has been reached, you may wish to establish time limits, or some record-keeping strategy.
10. Hold up cards on which numbers less than 19 are written. Ask the child to name pairs of addends for each number shown. For example, if you hold up

a card on which 7 is written, the child should respond "five plus two," or "six plus one," or the like. Vary the activity by challenging the child to name all the appropriate combinations.

11. Use two cubes made from sponge rubber. On each face of each cube write or paste single-digit numerals. These can be any single-digit numbers forming one of the addends of combinations on which the children need practice. The children take turns throwing the cubes and naming aloud the sum of the two digits that are face up when the cubes land.
12. When conducting drill and practice exercises with children make every effort to circumvent opportunities for the child to rehearse incorrect responses. Until you are fairly sure that the child knows the sums, allow the child to use some means of determining the sums for those expressions of which he or she is not sure. Some children like to use chips; some children like to use tally marks. Others can be taught to use an addition chart. Such a chart can be copied onto a piece of cardboard and kept inside the child's desk, or the chart can be taped to one corner of the top of the desk. You can encourage the children to avoid undue dependence on such aids by suggesting that the learners keep a record of the time needed to complete independent practice sheets. As the children become more proficient in remembering the combinations, this time will decrease. Set a goal for each child and encourage the child to work toward his or her goal.

The number of ways in which we can do an addition example are many. Some are perhaps more efficient than others, some may occupy less space on a sheet of paper than others, some may be more in tune with our level of comfort and understanding than others. None, however, is correct to the exclusion of others.

Most of us probably learned the conventional algorithm for addition when we were in grade school. This algorithm directs us to write the addends one under the other, using care that the digits representing the same powers of ten are lined up one under the other in a column, as in example A in Figure 6–2. We then find the sum of the numbers in each column, working from right to left. The completed example would look like example B in the illustration. Because no sum in a column exceeded 9, no regrouping (carrying) was necessary.

Let us look at another example, example C in the illustration. The sum of the digits in the first column is fifteen ones. The conventional algorithm directs us to rename fifteen ones as one ten and five ones, to write the 5 under the ones column, and to add the one ten to the digits in the tens column. Since no more regrouping is necessary, the completed example would look like example D. If the sum of the numbers in the second column exceeded 9, we would rename the sum as n hundreds and m ones, and add the n hundreds to the third column. A similar procedure is followed every time regrouping is necessary. Some children find it

Figure 6–2 The Conventional Addition Algorithm

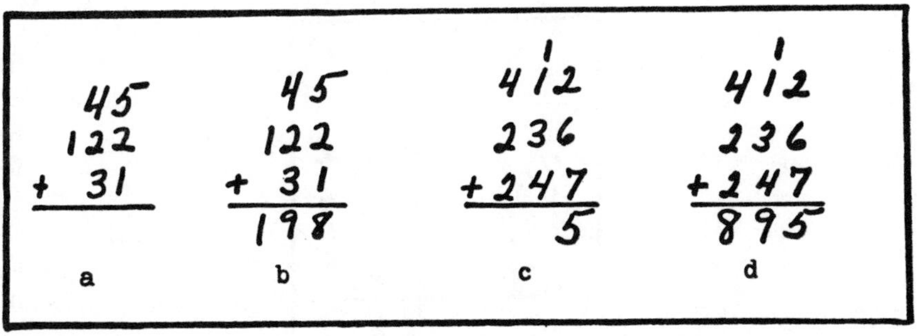

a	b	c	d
		1	1
45	45	412	412
122	122	236	236
+ 31	+ 31	+247	+247
	198	5	895

easier to see the reason for the "carried number" if they write it at the bottom of the appropriate column, rather than at the top.

Finding a sum is as easily done from left to right as from right to left. The former direction has certain advantages. The first of these is the freshness one brings to adding those figures that represent higher powers of ten. The effects of fatigue or restlessness are left for the ones column. A second benefit is derived from the fact that one carries or regroups in the "answer." For many children this process affords a better understanding of what regrouping is about. Finally, some children have developed a flair for adding from left to right. For these children it is sometimes easier to learn the correct left-to-right procedure than to begin from the beginning to teach the more traditional right-to-left strategy. The left-to-right procedure is demonstrated in Figure 6–3. Children who add from left to right should be encouraged to rewrite the sum so that all the digits are on a straight line.

Figure 6–3 A Left-to-Right Algorithm for Addition

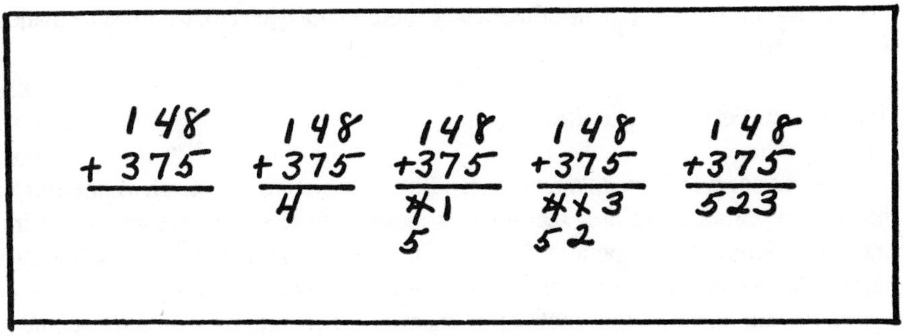

148	148	148	148	148
+ 375	+375	+375	+375	+375
	4	~~4~~1	~~4~~~~1~~3	523
		5	52	

Figure 6–4 A Modified Algorithm for Addition

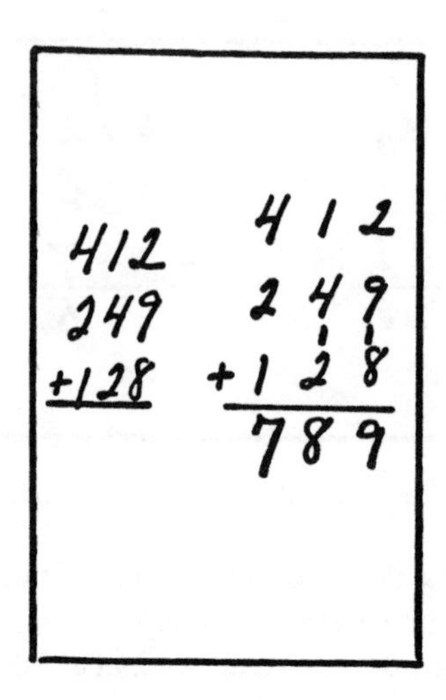

For children who have trouble adding long columns of numbers, a modification of the conventional algorithm can help. In this method, the partial sums are written down as soon as 10 or more is reached. To use this procedure the child should write or copy an example so that plenty of space is left between addends, since space is necessary for the digits of the partial sums. When regrouping is necessary several times in an example, this procedure requires the writing of many extra digits. For this reason, children using the method may need help in keeping track of the partial sums. Crossing out digits that have already been added is one way to do this. Figure 6–4 presents an example of this partial sums algorithm.

Alternate algorithms need not be taught as a matter of course. In discussing them our purpose is to provide the classroom teacher with options in teaching computational procedures to learning-disabled children. Whether to stay with a conventional strategy or to introduce a new method is a decision that must be made by the instructor.

Whole Numbers, Subtraction

Children seem to come to an understanding of subtraction as much as one and a half to two years after gaining facility with addition (NAEP, 1970), even though in most programs the two operations are taught almost simultaneously. That finding argues the need for much practice if subtraction is to be mastered.

We use the operation of subtraction in many different situations. *Remainder subtraction* is the term used to signify that the operation is being used to find out what remains in a set after a subset has been removed. *Comparative subtraction* denotes the use of the operation to find out by what magnitude one quantity exceeds another. *Additive subtraction* indicates that the operation is being used to find out how much must be added to one number to equal another number. Each type of problem calls for a different thought process, but the operation is the same in all cases. Remainder subtraction is the easiest to demonstrate by way of manipulatives. Learning-disabled children must have a great deal of experience applying subtraction in all three situations.

Since remainder subtraction involves the existence of a beginning set, representations of items in a set form a natural first step in instruction. The number of elements in the set is written down as the top number in the example. The number of items to be removed from the original set is represented by the bottom number, and the number of items remaining is the answer. The action sequence with manipulatives is this: (1) represent the larger set, (2) remove the necessary number of items from the larger set, and (3) count the remaining items to find out what is left.

To represent remainder subtraction on the number line, locate the top number first. Represent the bottom number by taking steps "backward" on the number line, each step equal to one unit on the number line. The number located at the end of the last step is the answer. Counting steps, rather than counting points on the number line, seems to be more effective in avoiding the question "Do I count the first point?"

To illustrate comparative subtraction, use small items to show the two sets under consideration. Lay out both sets in a straight line, one under the other, so that members of one set can be paired with members of the other. To find the answer to the comparative question, pair the members of one set with the members of the other set until no more pairs can be formed. The set in which there are elements without partners is the larger set. The magnitude of the difference can be found by counting the items without partners.

The representation of the third application of subtraction demands the awareness of the hypothetical larger set. Some children do this naturally in their finger counting for subtraction. For example, when asked to find the answer to an expression such as $9 - 6$, some children will hold up six fingers and then continue counting from six, holding up the necessary fingers, until nine is reached. They derive the answer by noting how many extra fingers were needed. In a very real sense, this is an understanding of additive subtraction.

When larger numbers are required, other means of visualizing the set having the desired cardinality must be used. The child can locate the number symbolizing the cardinality of the desired set and then the number representing the cardinality of the set that exists. By counting the number of steps from one number to the other the child can arrive at the answer. Experiences such as these may help a child to understand subtraction as the inverse of addition.

Many of the activities suggested for drill and practice on the addition combinations are useful for helping children memorize the subtraction combinations. We will mention here only activities that, for one reason or another, are particularly useful for drill and practice in the area of subtraction.

1. An addition chart may help some children who have not as yet memorized the subtraction facts. The child is to locate the bottom number of the example in the first column to the left. The child should then move his or her finger across that row until he or she finds the top number. By moving his or her finger up that column to the very top, the child will find the answer. If the child must learn the facts as they are written horizontally, the directions can be adapted appropriately.
2. Prepare a list of sets of three numbers. The numbers in each set should be such that they can be rearranged to make a complete subtraction sentence. The children's task is to write two subtraction sentences for each set of three numbers. For example, for the set 2, 6, 4, the children should respond $6 - 2 = 4$ and $6 - 4 = 2$.
3. Prepare with the children two decks of cards. On each card in one deck write each of the numbers from 10 to 18, one number to a card. In the same fashion write the numbers 0 to 9 on the cards in the other deck. Place both decks face down, next to each other. A child picks one card from each deck and states the difference, or writes the complete subtraction sentence.

We shall discuss some of the better known alternative algorithms for subtraction and try to illustrate their potential usefulness for children with learning disabilities.

The conventional algorithm is sometimes called the decomposition algorithm because of its use of the equality between one group of ten and ten ones, one group of one hundred and ten groups of ten, and so on. Many of us learned to call the notation for this process *borrowing*. At present the more commonly used term is *renaming*. That is, we rename one ten as ten ones; we do not "borrow" from the tens place to add ten ones to the top number in the ones column.

In teaching this algorithm we make use of the understanding of place value discussed earlier. Using manipulatives, we can demonstrate the decomposition process; using symbols we can "write down the story" of what is "happening" during the manipulation. Because the symbolic expression is so closely related to the procedure used in manipulation, this decomposition algorithm can be and should be easily taught.

A second subtraction algorithm, and one quite widely used, is the equal additions algorithm. The algorithm makes use of a law of algebra that, inelegantly stated, says that adding the same number to the top number and the bottom number in a subtraction example leaves the answer unchanged. The algorithm makes use of the equality between one ten and ten ones, one hundred and ten tens, and so on. The procedure is shown in Figure 6–5.

Figure 6–5 The Equal Additions Algorithm for Subtraction

52 − 37	Original example
5,2 −4̶3̶ 7	Add ten ones to top number Add one ten to bottom number

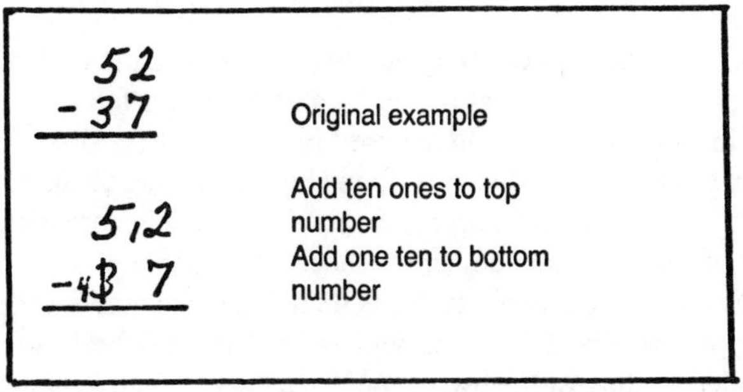

This method has both advantages and disadvantages. It does not lend itself to demonstration by way of manipulatives easily. On the other hand, examples like 400 − 134 and 1000 − 475 are easier because the differences in each column can be found one right after the other. To use the conventional algorithm, one must first rename/"borrow" several times before any subtraction can be performed.

The advantages cited for left-to-right addition present themselves in left-to-right subtraction also. Once again all renaming is done in the "answer," not in the numbers above the line. The examples shown in Figure 6–6 demonstrate the procedure for left-to-right subtraction. Again, the answer should be written on a straight line.

Whole Numbers, Multiplication

We think of multiplication as a shortcut to finding the sum of the cardinal properties of several equivalent sets. Thus, if every seat in every row of a theater

Figure 6–6 A Left-to-Right Algorithm for Subtraction

621 − 357	621 −357 3	6,21 −357 3̶7 2	6,2,1 −357 3̶7̶4 26	621 −357 264

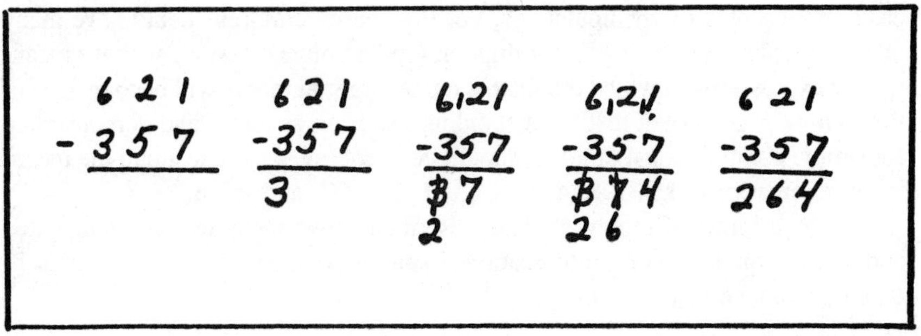

were occupied, and if each row had the same number of seats, we would multiply the number of rows by the number of seats in each row to find the number of people in the theater.

We can also use multiplication in another class of situations. Imagine yourself going on a shopping spree for clothing. You find and buy four pairs of slacks and three shirts that can be mixed and matched in every way possible. The number of matches you can make will always be found by multiplying the number of items in the first set by the number of items in the second set, that is, for the example given, the number of shirts by the number of slacks. It is such a process of mixing and matching the members of two sets that is involved in the mathematical definition. Children with learning disabilities will benefit from experiences designed to promote understanding of the use of multiplication in both situations.

Multiplication as repeated addition can be demonstrated in several ways. Equivalent sets can be represented through the use of small items. The child should be encouraged to count first the number of sets and then the number of items in each set. Only after the learner is aware of these two numbers should the task of counting all the items or adding the number of items in each set be assigned.

A second representation of multiplication makes use of rectangular arrays. In this model, the expression 3×4 would be represented as three rows of four items each. Once again, the child should count the number of rows first and then the number of items in each row. After these two numbers are obtained, the child can find the total number of items by counting or adding. The suggestion that the child be made aware of the number of groups and the number of items in each group in the first representation and the number of rows and the number in each row for the second representation stems from the relationship between the two numbers as the factors in a multiplication expression. The expression 3×4 can be interpreted as meaning three groups of four or three rows with four items in each row. This is a meaningful interpretation because it can be linked to the practical situations in which we most frequently use multiplication to find a solution to a problem.

The third means of representing a multiplication expression makes use of a number line. The expression 3×4 on a number line would be represented by three jumps of four units each.

The conventional algorithm for multiplication is somewhat removed from being easily represented by manipulatives. For this reason children should have many experiences in representing single-digit multiplication expressions so that a meaning for expressions in which the factors name larger numbers will become part of their knowledge. When that understanding has been acquired and demonstrated for simple combinations, children should be encouraged to state aloud the meaning of expressions like 32×4, 5×15, 143×27, and so on.

The three forms of representations described above make use of the repeated addition interpretation of multiplication. Examples of the three types of representation are shown in Figure 6–7.

Figure 6–7 Three Representations of Multiplication

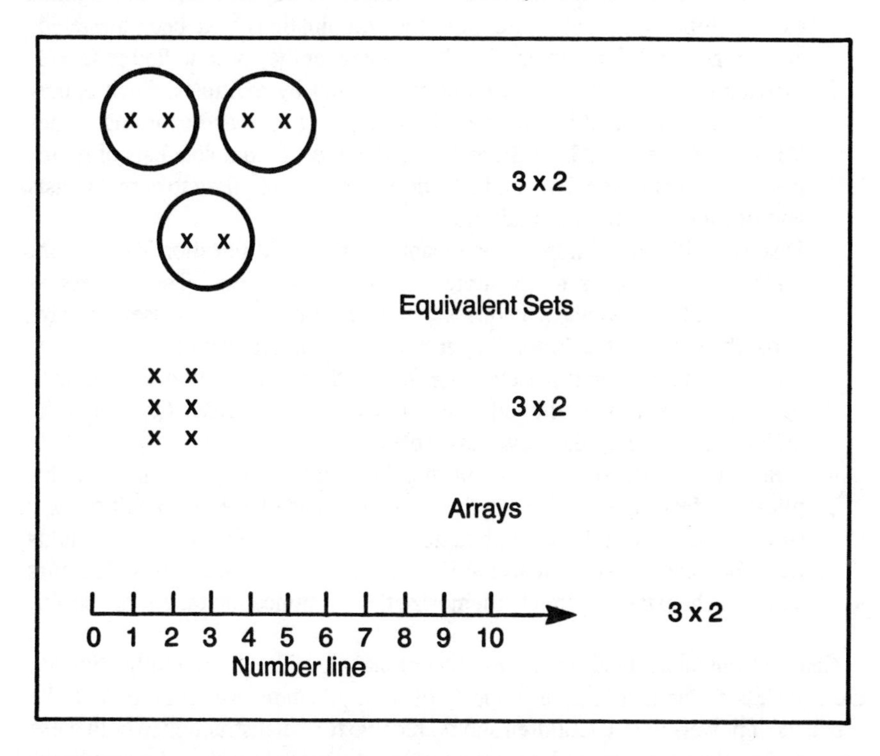

The cross-product interpretation of multiplication, explained earlier as the mix-and-match interpretation, can be translated into a fourth way of representing multiplication expressions. Each factor is represented by a number of strips. One set of strips is placed horizontally; the other is placed vertically over the horizontal arrangement. The expression 3×4 would be represented by criss-crossing three strips and four strips. The product is found by counting the number of intersections. There is a certain economy to be found in this approach. The number of items to be arranged is fewer than those required by the equivalent sets model, or the array. Another advantage is the potential use of the model for representing expressions in which the factors are two- or three-digit numbers. By varying the color or the width of the strips, tens and hundreds can be represented.

The following activities are designed to help children develop an accurate understanding of multiplication and to memorize the multiplication combinations.

1. Help the children prepare a multiplication chart. Discuss some of the patterns found on the chart: the symmetry of the body of the chart, the

endings of the multiples of 5 and the multiples of 10, the sum of the digits in each multiple of 9, and so on. Until the combinations have been mastered, encourage the children to use the chart whenever they need to find products.
2. Provide opportunities for the children to count by multiples. Such opportunities can be found in games such as skipping rope and counting games ("Hide and Seek," "Giant Steps," and the like). Incomplete lists of multiples can be displayed daily and the children can be required to copy the lists and to fill in the missing multiples.
3. Discuss with the children the meaning of a multiplication expression, making up examples to illustrate different expressions. The expression 2×5 might represent two houses with five rooms in each house. If we reverse the order of the factors to get 5×2, the interpretation becomes five houses with two rooms in each house, hardly the ideal situation, even though the number of rooms altogether is the same in each case. Challenge the children to make up their own examples.
4. Strategies for effecting and monitoring the memorization of the basic multiplication facts are similar to those recommended for the addition and subtraction facts. If those combinations have been taught using the methods described earlier, some learning should have taken place. Those learning skills can help the children to memorize the multiplication facts more easily.

Central to an understanding of the conventional algorithm for multiplication are the concepts of the distributive property of multiplication over addition and the products of powers of ten. Children should receive formal instruction in both these areas before they are required to master the traditional algorithm. From a functional point of view, we seldom experience the need to find products when each factor consists of more than three digits. There is little justification for requesting children to spend time on lengthier examples.

The conventional algorithms for multiplication can be very difficult for children with learning disabilities. Most of the reasons for the difficulty lie in the algorithm itself. The computation of a product for the factors 48 and 259 requires the child to remember six of the basic combinations, to remember the products of six different multiples of powers of ten, to write anywhere from 13 to 17 digits in their appropriate places, to remember when to add and when to multiply, and to remember five of the basic addition combinations and the sum of at least four pairs of addends that are not found in the basic combinations. The attainment of all these skills and their appropriate uses represents a very formidable task.

An algorithm that does much to aid children in organizing the placement of the digits while reducing the need to "carry" during the multiplication process is the diagonal lattice procedure. The approach is demonstrated in Figure 6–8. The numbers in the body of the matrix are the products of the numbers above and the numbers to the right. These numbers are then added along the diagonals of the

Figure 6–8 Diagonal Lattice Procedure for Multiplication

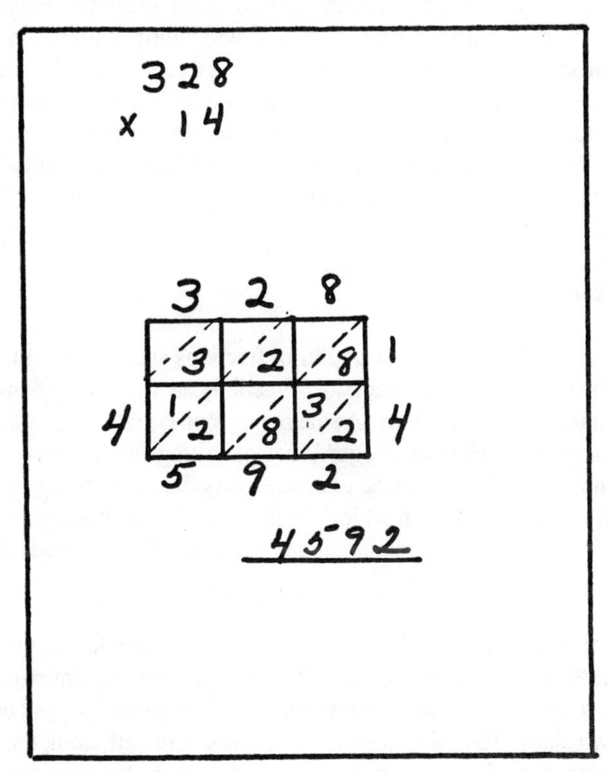

matrix. The algorithm is characterized by a different procedure for writing the story of what happens when we multiply, rather than a gimmick.

Whole Numbers, Division

In grades 2 through 4, experiences that emphasize the meaning of division are essential. There are two situations in which division is the appropriate operation. The first of these is called measurement division: "A decorator has 24 yards of drapery material and an order for pairs of drapes each of which requires four yards of material. How many pairs of drapes can the decorator make with the material at hand?" The second use of division is usually labeled partitive division: "A decorator has 24 yards of cloth on hand and an order for eight cushion covers. How many yards per cushion cover can the decorator use?" In the first case, the translation of the problem into a number sentence would result in "some number times 4 equals 24. In the second case, the number sentence would read thus, "8 times some number equals 24." In measurement division the first factor is the unknown quantity; in partitive division the second factor is the unknown.

Because of the relationship between division and multiplication, the same type of representations can be used. Instead of counting to determine the total number of elements involved, we count to find the number of groups or the number in each group. Thus, to represent 12/3, we might count out sets of three until all 12 items have been used, or we can "deal" one item to each of three sets until all 12 items have been used. Hence, many of the activities suggested for multiplication can be adapted for division.

Fractions, Numeration

During grades 2 through 4, instruction in fractions should concentrate on the meaning of the symbols we call fractions. Fractions, as we usually think of them, are a subset of the rational numbers. As the name implies, rational numbers are the result of forming the ratio of two integers, that is, writing them, one over the other, in the form m/n. These ratios can be used to signify the relationship of a part to a whole, and that is the context in which we usually think of fractions.

Activities for building an understanding of fractions should include the following.

1. Over a period of time, work with the children to prepare fraction kits, one for each child. Begin by providing each child with two square regions. Each child is to cut one square region in half on a line parallel to and equidistant from two sides. The other region should be cut in half along the diagonal. Each piece should be labeled with the symbol ½ on one side, and placed in an envelope. Repeat the activity for circular regions and for rectangular regions that are not square. Move on to representations for thirds, fourths, fifths, sixths, eighths, and tenths. As each new fraction is represented, relate it to previous fractions. Thus, when fourths are completed, compare the fourths with the halves to develop the relationship between two fourths and one half.
2. Using the fraction pieces described above, develop fraction families consisting of members of equivalence classes: different names for one half, different names for one third, different names for two thirds, and so on. Children can make a notebook or a storybook about each new equivalence class.
3. Assign children to work together in pairs with their fraction kits to develop an understanding of improper fractions and mixed numerals. After some skill in this area is attained, encourage children to find and to demonstrate with their fraction kits different names for such numbers as 2½. Solutions should include 5/2 and 1 and 3/2.

4. On a worksheet, draw representations of different fractions. Ask the children to use their fraction kits to represent the same fractions in different ways.
5. List pairs of fractions on the chalkboard. Both fractions in each pair should have the same denominator, or the same numerator. Ask the children to use their fraction pieces to determine which of the two fractions names the larger part. Increase the number of fractions in each set to three. Repeat the activity. Through induction the children should be able to derive the rules for comparing fractions with like denominators and for comparing fractions with like numerators.
6. Using their fraction pieces, children can represent simple addition examples and simple subtraction examples.

Geometry, Shapes

Instruction in geometry deserves more time and attention in grades 2 through 4 than it often receives. That is even more true for learning-disabled children. Many children are more able to handle geometry concepts during those years than the demands that arithmetic makes upon them.

Lines and paths, two- and three-dimensional shapes, can be the context for some experimentation and exercise in inductive thinking. At the same time the children can learn valuable material, both from a mathematical perspective and from a vocational standpoint. The activities suggested below are just a sample of the many that a good mathematics program will provide for children at this level.

1. Explain through demonstration the meaning of straight paths, curved paths, and broken paths. Point out the models of the letters of the alphabet (print) or the models of the numerals 0 through 9. Ask the children to identify examples of straight paths, curved paths, broken paths.
2. Ask the children to demonstrate by walking one of the paths under discussion: straight, curved, broken. The other children are to draw pictures of the kind of path that is being demonstrated.
3. Point out again the models of the letters of the alphabet (print or script). Ask the children to identify examples of open paths and closed paths.
4. Provide the children with worksheets on which are drawn examples of triangles and rectangles. Ask the children to use crayon of one color to color the inside of the triangles and to use crayon of another color to color the inside of the rectangles. Repeat the activity for other combinations of polygons. At a later time, prepare worksheets on which are drawn examples of squares and examples of rectangles that are not squares. Ask the children to distinguish between the two.

5. Give the children riddles to solve such as the following: "I am a shape that has four sides, all of which are equal. I also have four right angles. What am I?"
6. Provide the children with models for 90° angles, 30° angles, 60° angles, and 45° angles. These models can be wedged-shaped pieces of tagboard. Provide the children with drawings of the same angles. Ask the children to match the models with the drawings. At a later time, give the children worksheets on which are drawn examples of 30°–60° right triangles and 45° right triangles. Ask the children to use the models to distinguish between the two types of right triangles. After the children have learned to measure length to the inch, prepare drawings of the two types of triangles so that the sides of the 45° right triangles that form the right angle are equal and measure one, two, or three inches. Ask the children to measure the sides of those right triangles, and to arrive at a rule about the lengths of the sides of a right triangle in which the two sides that form the right angle are equal. Repeat the activity for the 30°–60° right triangle. In this case, draw the triangles so that the side opposite the 30° angle is one inch, two inches, or three inches. The children should measure that side and the side opposite the right angle and derive the rule from those measures. In addition to acquainting the children with two special and important right triangles, these activities provide practice in measurement and in following directions.
7. Models like the one shown in Figure 6–9 can be used to provide the children with experience in constructing three-dimensional models from two-dimensional drawings. The completed three-dimensional figures can be used for many exploratory activities: comparing the three-dimensional shapes one to the other; counting and keeping track of the number of vertices, edges, and faces; finding everyday items that have the same shapes, and so on.

Geometry, Relationships

The study of relationships in geometry can be a worthwhile experience for the learning-disabled child. Noting relationships of congruence, symmetry, and similarity demands attention to detail, discrimination, and the ability to see likeness in spite of a change in position. Many learning-disabled children need these experiences.

1. Prepare a worksheet on which are drawn triangles of the same shape (that is, all triangles should be equilateral, or 30°–60° right triangles, or the like) but of different sizes. Ask the children to color in all the triangles of one size with one color, the triangles of another size with another color, and the triangles of yet another size with a third color.

Figure 6–9 Pattern for a Model of a Triangular Prism

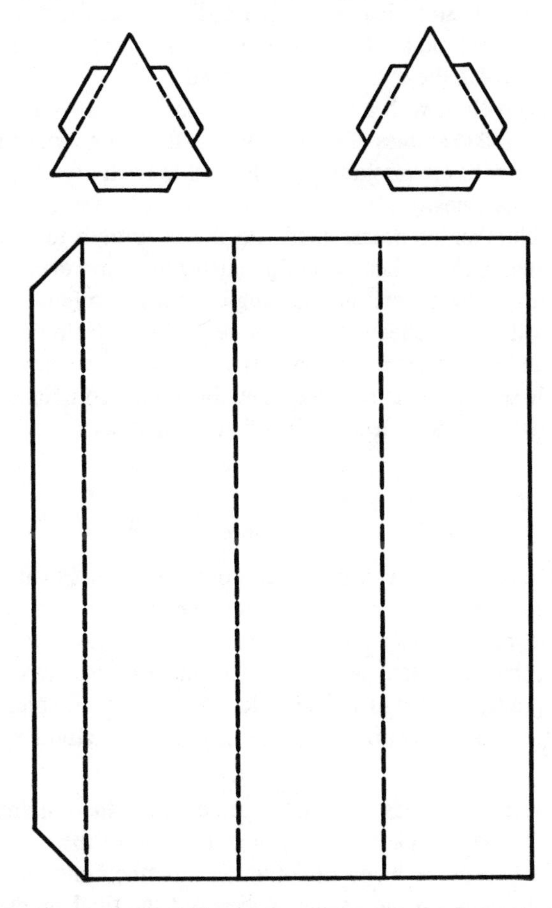

2. Prepare a worksheet on which two columns of shapes are drawn. Each shape in the second column should be similar to, but not congruent to, a shape in the first column. Ask the children to draw a path from each shape in the first column to the shape most like it in the second column.
3. Provide the children with worksheets on which are drawn different shapes. Ask the children to cut out the shapes. When all shapes have been cut out, ask the children to determine which of the shapes have a line of symmetry. Encourage the children to fold the shapes to confirm their guesses. At a later time, again provide the children with worksheets on which shapes are drawn. This time ask the children to identify shapes having a line of symmetry without cutting the shapes out.

4. Ask the children to collect items that demonstrate the concept of symmetry. An orange slice, some leaves, a tomato slice, some tile patterns, all afford examples of symmetry. Help the children to prepare a bulletin board or collage on which the collection is mounted.
5. Provide each child with a few pieces of masking tape on a ruler. Ask the children to find examples of parallel lines in the classroom and to mark parts of the lines with the masking tape. Repeat the activity at another time for perpendicular lines.
6. Invite a plumber or a carpenter into the classroom to discuss with the children the many applications of geometry in their work.
7. Explain and demonstrate the meaning of translation (slide), rotation (turn about a point) and reflection (flip) to the children. Help the children make simple hand puppets representing different geometric shapes. The children can use these puppets to put on a show illustrating combinations of translations, rotations, and reflections in a dance context.

Measurement, Metric Geometry

For children entering into concrete operations, experiences with measurement afford the physical interaction with the environment that can be most helpful to the development of the necessary cognitive structures. As the scope and sequence chart indicates, there are many areas in which such experiences can be provided for children in grades 2 through 4. The topics of length, area, volume, and angles supply the context for many different types of hands-on activities.

1. Using centimeter cubes that link together, help the children construct a model of a meter stick by linking together 100 of the cubes. This group activity will provide many opportunities for the children to discuss the relationship between the smaller cubes and the final product. While the resulting meter stick is not rigid enough for any practical use as a measuring stick, it will have served its purpose as an introduction to the relationship between the centimeter and the meter. Inch cubes can also be lined up to illustrate 36 inches to the yard.
2. Collect a number of large blocks, books, or empty boxes. Using some of the materials, build a tower on top of a table. Ask the children to build a tower as tall as yours on the floor. Observe the children as they work. Do they attempt to build a tower whose top is as high as the top of yours? Do they employ a guessing technique? Do they employ a tool to measure the height of your tower before proceeding with the task? The different strategies indicate different levels of cognitive development, and consequently different needs for instruction.

3. Provide the children with models of different geometric regions and different tools that could be used as units of nonstandard measures: pieces of string, unit squares, narrow sticks, tiles, and the like. Ask the children to measure the area of the regions and the length of the boundaries of the regions (the perimeter). Observe the accuracy with which the children select measuring units.
4. Invite into the classroom an employee of a carpet store or a wallpaper store. Ask him or her to explain to the children the different ways in which measurement plays a role in home decorating.
5. Provide models of two rectangular regions, both longer than they are wide. Place one region so that the longer sides are in a vertical position; place the other so that the longer sides are in a horizontal position. Use the models to provide the background for a discussion of height and width, and the dependence of either concept on orientation.

Measurement, Time

Time is a most abstract quantity. Children need to experience this dimension of our existence in many ways in order to develop an awareness of the meaning of time. A remedial teacher tells the story of an eight-year-old boy she was tutoring at home. At the end of a reading lesson, the teacher suggested that the child work on a practice exercise for five minutes before going out to play. The child burst into tears. The teacher was surprised, and turned to the mother for an explanation. The mother explained that the boy had no notion of what five minutes was, and thought it was a very long time. The story reminds us that we must be careful in our use of terms relating to time when we are dealing with children.

1. Discuss with the children some of the nonstandard units they have invented for dealing with time. One young girl used to measure distance on a long trip by figuring out the number of "Muppet" shows she could watch during the time it took to make the trip. Other children measure the length of a commercial television program by using the number of advertisements that occur regularly during the show. If you can encourage the children to share this kind of information with you, you will have an inkling as to how sophisticated their concepts of time are.
2. If possible, obtain a model of the inner workings of a large clock that tells time through the use of rotating hands. Point out the different gears to the children, explaining how they work to move the hands of the clock.
3. Obtain a model of our solar system illustrating the relative positions of the sun, the moon, and the earth. Discuss with the children the role these bodies play in our measurement of time, demonstrating the movement of the different bodies.

4. Write the names of several holidays on the chalkboard at random. Ask the children to list the holidays in the order in which they appear in a calendar year. Evaluate the lists to determine how well the children are able to sequence those events.

SUMMARY

In the preceding pages we have outlined the rudiments of a scope and sequence for learning-disabled children in grades 2 through 4, and we have listed a number of activities that the teacher of learning-disabled children will find helpful in providing quality mathematics instruction for these children. The activities are merely a sample of the many types of experiences that will enable children to better understand and achieve in mathematics.

REFERENCES

Brigance, A. (1977). *Brigance diagnostic inventory of basic skills.* Woburn, Mass.: Curriculum Associates.

Cawley, J.F., Goodstein, H.A., Fitzmaurice, A.M., Lepore, A., Sedlak, R., & Althaus, V. (1976). *Project MATH, Levels I and II.* Tulsa, Okla.: Educational Progress Corp.

National Assessment of Educational Progress, (1970). Washington, D.C.: U.S. Government Printing Office.

Chapter 7

Curriculum and Instructional Activities The Upper Grades

Robert A. Shaw

The upper grades represent a significant growth and development period for learners. It is during this time that the basic concepts and skills of mathematics must be presented and learned in a meaningful manner. Understandings and attitudes developed while learners are in these grades serve to enhance or hinder more learning of mathematics. The importance of this span of grades was recognized by the committees involved in the development and organization of the National Assessment of Educational Progress (NAEP) when they selected grades 4, 8, and 11 from which to select samples of learners. They were (and still are) interested in what concepts and skills the learners at these grade levels possess.

The question, "What does the student possess in terms of mathematical concepts and skills" serves as a beginning for this chapter. From this point a task-analysis procedure will be used to explore what should be taught and how it should be taught to learning-disabled youth. The chapter will conclude with selected content-strategy-method examples.

ENTERING CONCEPTS AND SKILLS

When entering the fourth grade what do the learners possess in terms of concepts and skills of mathematics? The importance of this question cannot be overemphasized, although, unfortunately, in many schools, the question is not even asked. In such schools the program in mathematics is the same for everyone, as is the instruction. If effective planning is to occur this question must be answered, especially when we consider individuals with learning disabilities. It is at this fourth grade level that we have the opportunity to correct developmental mathematics "gaps" and to prevent other such gaps from occurring if we select the appropriate instructional content and materials and use the content and materials in a meaningful way.

How do we determine the current status of learners' understanding of mathematics? While the topics of assessment and evaluation are considered elsewhere in this book, some basic ideas should be presented here. First of all, the results of the norm-referenced and domain-referenced instruments that are given in many schools on a routine basis are readily available sources of information. (Data from such instruments are often called nomothetic data.) From such instruments we get summative information, scores that indicate where an individual is in regard to his or her total group at that particular grade level at that particular time. Grade-level equivalency scores or percentiles are often reported and from these scores many conclusions are drawn.

"John is at grade level in terms of his achievement in mathematics." "Betty is on grade level in terms of mathematical concepts and skills." What do these statements mean? If John and Betty are in the same mathematics class in the same school, does this mean that they possess the same mathematics concepts and skills at the same degree of understanding? Equivalent scores can be obtained by many different combinations of right answers. John may know nothing about fractions but does well in other areas of the measuring instrument, while Betty may have difficulty with subtraction of whole numbers but may understand fractions completely. John and Betty could obtain the same total score, but they have different levels of understanding of mathematics. From such examples we can conclude that many instruments give us only global information in terms of performance of individual students. Based on these instruments the learners are either *on* grade, *above* grade, or *below* grade level, but we do not know the specific details. What we have is an increased awareness of possible content areas of difficulty.

When we need a more detailed analysis of entering concepts and skills, a status assessment is necessary. Such an assessment involves both summative and formative evaluation procedures. Any available records of grades, teacher recommendations, checklists of concepts mastered, tests and scale results, and observation reports should be considered in devising plans for instruction. We need to establish a beginning and to get some direction from available information. Such norm-referenced tests as the *Bobbs-Merrill Arithmetic Achievement Tests* (Kline & Baker, 1963) may be used to show general achievement patterns from grades 1 to 4. Domain-referenced tests such as the fourth grade test of the National Assessment of Educational Progress in Mathematics may be used to explore strengths and weaknesses in goal and objective areas of mathematics. The *Diagnostic Mathematics Inventory* (DMI) (1975) of CBT/McGraw-Hill may be used as a criterion-referenced instrument to obtain individual item data on each learner. More detailed information may be obtained from such instruments as the *Inventories of Project MATH* (1976) and the *Buswell-John Diagnostic Test for the Fundamental Processes in Arithmetic* (1925). (See Chapter 3 for an in-depth discussion of assessment.)

From a combined testing package such as this we have an indication of strengths and weaknesses, thus, a general set of guidelines or pattern of symptoms (Underhill, Uprichard, & Heddens, 1980) is now available. Such information may be used for writing objectives, determining intraclass groupings, identifying areas for individual help or involvement of a workmate, establishing guidelines for differentiated assignments and quizzes, and initiating clinical interviews for more detailed diagnostic procedures.

It should be noted that in attempting to determine what learners possess we have no idea (unless a longitudinal study is underway) what the learners know from their past experiences or what they learned directly or indirectly from formal instruction; however, to determine the mathematical content of the curriculum we will pursue our task analysis through the learning question and assume that there is a relationship between what students learn from instruction and what they are able to demonstrate. The influencing factors of abilities, interests, levels of motivation, etc., will be considered below along with strategies and methods.

DETERMINING CONTENT

Two possible paths lead to the present knowledge and skills of the learners: (1) what was presented by the instructor and (2) how the content was presented. This section is concerned with the *what* question. Several related questions begin to evolve as the task-analysis procedure is undertaken. How do we know what was presented? Are weaknesses in the backgrounds of the learners a result of nonlearning of presented materials or nonteaching of critical concepts and skills? What is the relationship between what is presented and what is learned? For example, a group of teachers of sixth grade mathematics was given a scope and sequence chart for the sixth grade content. This group was asked to indicate what topics from the chart were formally presented in the mathematics classroom. Seventh grade teachers were asked to indicate from the same scope and sequence chart for the same group of learners what entering concepts and skills the learners had when they came to them. The sixth grade teachers indicated that 96 percent of the topics from the chart had been taught. The seventh grade teachers indicated that general knowledge covered only 6 percent of the topics. Achievement test data at the end of the sixth grade gave results between the two extremes.

Other questions emerge as we continue the analysis. What was planned by the instructor? What content was outlined for the instructor to present? What amount of the planned content was presented? What amount was learned? What content is of most worth for the learners in grades 4 through 8? That is the crucial question. Input comes from mathematics educators and special educators, content specialists, psychologists, publishers, and society in general. Such information appears in articles from professional journals to newspapers, reports and recommendations from committees, and sample and experimental program materials.

Considerable time and effort have been expended in attempting to decide *what* to teach. In special education the works of Inskeep (1938), Ingram (1968), and Meyer (1972) reflected this concern for what to teach from the 30s through the 60s. Serious concerns about mathematics began in the 50s and this was followed by a content revolution in the 60s and 70s that called for (1) more understanding of the process of mathematics, (2) an expanding scope for the curriculum, and (3) exemplary programs. The report from the Cambridge Conference on School Mathematics (1963) contained the recommendation that the goals of elementary school mathematics should be to develop a familiarity with the real number system and the main ideas of geometry, and grades 7 and 8 should be devoted to the study of algebra and probability or algebra, geometry, and probability. We were to gain three years in content development.

Other recommendations came from the National Council of Teachers of Mathematics (NCTM, 1959) and the College Entrance Examination Board (1959). The secondary level (grades 7–12) received the major emphasis and financial support for new programs. Gradually, the new materials began to filter into the mathematics classroom; however, this filtering process was (and remains) very slow. According to the National Assessment of Educational Progress (NAEP) in mathematics only 16 of 200 items given to thirteen-year-olds in 1972 and 1973 reflected content material that was not likely to have appeared prior to 1960. For nine-year-olds only a few items on sets and geometry indicated any changes (NACOME, 1975).

To compound the situation, as the developmental phases of the projects ended, attention was turned to the psychological implications of mathematics instruction. Efforts were made to match the levels of the new materials to the capabilities of the learners at different grade levels and variations increased tremendously. There is evidence to support the hypothesis that mathematics programs in our schools evolve in a direct relationship to the interests and abilities of those individuals in charge of such programs; thus, another reason for the variations.

Curricula for the less able students or students experiencing difficulty were almost nonexistent in the 60s and 70s. Early attempts include Braunfeed's (1969) *Stretchers and Shrinkers* and Phillips and Zwoyer's (1969) *Motion Geometry* for seventh and eighth grade courses and the NCTM (1970) booklets, *Experiences in Mathematical Ideas*. Other efforts in this area were given a boost with the passage of PL 94-142, The Education for All Handicapped Children Act of 1975. *Project MATH* and *Multi-Modal Mathematics* became the first large-scale projects in mathematics for children with handicaps to learning and achievement. These publications and their research provide the framework for this chapter.

Making the content selection and identification process still more complex is the back-to-basics movement. Attempts are being made to define basic skills. The

National Council of Teachers of Mathematics Position Statement on Basic Skills (1978) defines a broader scope than some of the national assessment data:

> The Council supports strong school programs that promote computational competence within a good mathematics program. . . . Computational skills in isolation are not enough, the student must know *when* as well as *how* . . . [S]tudents need more than arithmetic skill and understanding. They need to develop geometric intuition as an aid to problem solving. They must be able to interpret data. Without these and many other mathematical understandings, citizens are not mathematically functional. (p. 147)

The National Council of Supervisors of Mathematics (NCSM, 1978) has viewed basic mathematical skills as falling under ten areas: problem solving; applying mathematics to everyday situations; alertness to the reasonableness of results; estimation and approximation; appropriate computational skills; geometry; measurement; reading, interpreting, and constructing tables, charts, graphs; using mathematics to predict; and computer literacy.

Ross Taylor (1978) made a good distinction between skills and minimum competency: "*Basic skills* should be used as broadly as possible, including the wide range of mathematical skills people should have to function effectively in adult society. *Minimum competence* should be used as a relative term that has meaning only as it is used in a specific context (such as a graduation requirement)" (p. 89). From the instructional view there is a danger that the *minimum* may become *maximum*.

In regard to learning-disabled individuals, we should view basic skill areas as general qualities and organizational categories for providing experiences while keeping their needs in mind. *We must treat these learners no differently than others in terms of our expectations. We need to find the most effective means of instructing them.* Either the existing programs must be modified in terms of content selection or new programs must be developed and certain basic premises should precede either one of these activities to give specific direction.

- Most children with handicaps to learning and achievement are capable of learning more mathematics than is currently being learned by this population.
- Such an expanded knowledge is needed by these children.
- For many, if not most, of these children "specially designed instruction" should include more than simple exposure to a mathematics curriculum in the regular classroom.

To say that learning-disabled individuals are capable of learning more mathematics in no way defines what that content should be and to characterize the content as that which is necessary in everyday life helps in only a limited fashion; therefore, we must look to a variety of sources to determine the scope and sequence of our 4–8 program.

Content

In searching for major topics of mathematics in the upper grades in textbooks, curriculum guides, and committee reports, the following content areas (or strands) emerge. For grades 4, 5, and 6 the areas are (1) numeration and number systems that include the important place value concept, (2) whole number operations and properties, (3) geometry that does not include measurement, (4) operations and properties of fractions and decimals, and (5) measurement. Basic and unifying ideas such as relationships and patterns are included within the specific content areas. A theme underlying each area is problem solving. Figure 7–1 serves to illustrate a relational diagram for these five areas.

At other levels, the whole number section is expanded to include positive and negative numbers (integers); number theory elements such as evenness, oddness, lowest common multiple serve to connect whole numbers and fractions; ratio, proportion, and percents receive emphasis; and measurement is extended (see Figure 7–2).

In determining specific content, scope and sequence charts from various programs serve as our first source of information. This implies that the usual or most immediate source is the textbook designed for a particular grade level. Another question emerges. Are the editors of textbooks the real determiners of the curriculum? If the textbook is the only context for instruction, mathematical gaps will surely occur as a regular happening. Since the textbook demands a specific set of behaviors from the learners and some individuals may have difficulty functioning within these restrictions, the conditions are magnified for such learners.

Figure 7–1 Content Areas for Grades 4, 5, and 6

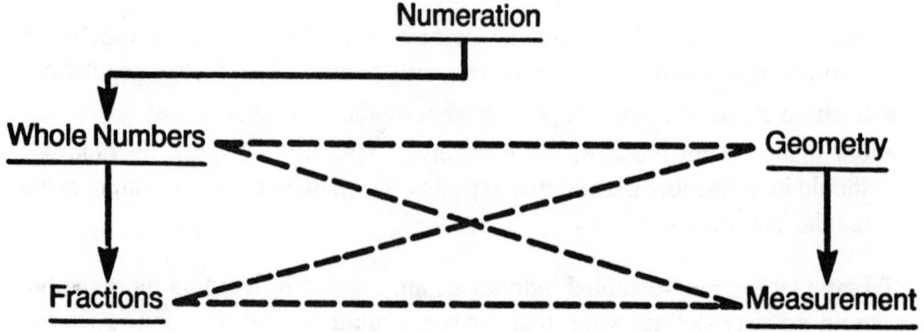

Figure 7–2 Content Areas

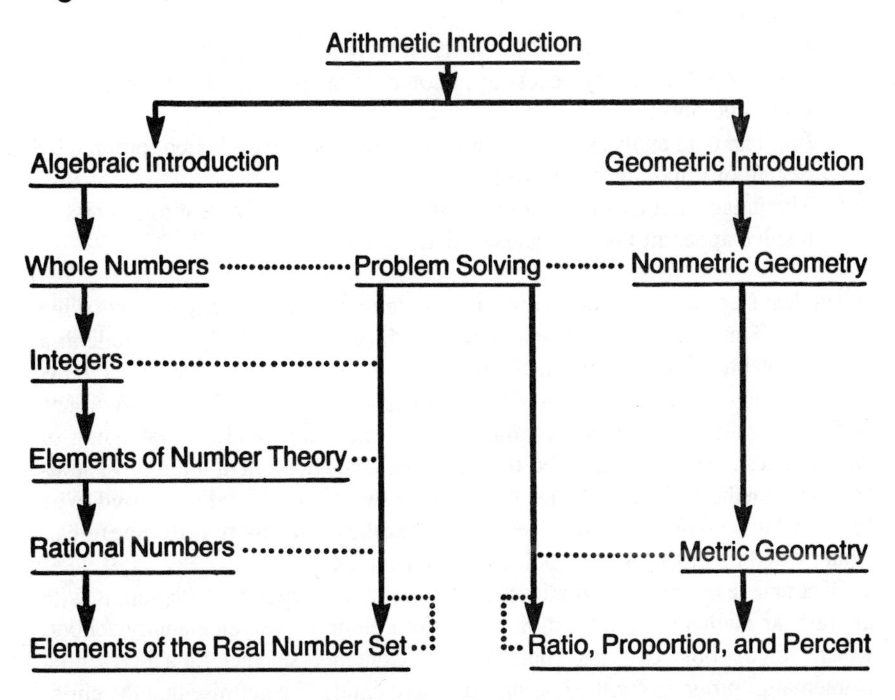

Farthest removed from the actual situation are the recommendations of various groups and committees as each group considers what *ought to be*, often without data to support its beliefs. In an attempt to bridge the gap between textbooks and recommendations from committees some states have prepared curriculum guides or guidelines that outline specific content sequences across the grade levels. The specific content represents (in many cases) a minimum core of concepts and skills and the danger exists, again, that the minimum may become the maximum.

In regard to mathematics for learning-disabled individuals a more realistic approach is to use a version of the Delphi technique and involve mathematics educators and special educators to select relevant content. Such an approach was followed by the developers of *Multi Modal Mathematics* (1980). Twelve judges were selected; six of these were individuals whose primary expertise was in the field of special education and the other six judges were mathematics educators. The following procedures were used:

1. Each judge was asked to list content appropriate for a learning-disabled student at the middle/junior high school level in each of four areas: geometry, measurement, fractions, and whole numbers.

2. An analysis of the twelve lists resulted in a composite list of topics for each of the four areas.
3. Each judge was then asked to rate each topic in each of the four areas on a scale from "1" (very necessary, should be taught) to "5" (unnecessary, should be omitted).
4. From an item analysis of the ratings a list of topics in each area, ordered by degree of importance, resulted.
5. The judges met as a group to discuss the final lists of content topics and to resolve apparent discrepancies or disagreements.

The last step proved to be the most enlightening for both the members of the project staff and the judges: the mathematics educators attributed to LD students a greater potential for learning mathematics than did the special educators. This greater confidence in the abilities of those students stemmed not from an ignorance of their academic problems but rather from a knowledge and understanding of mathematics content. Ultimately, this facilitated the adoption of alternative strategies and methods. The exchange between the two groups of judges proved to be beneficial to both sides; and it served to illustrate the great importance of the role of strategy and methodology in mathematics education.

Of course, existing mathematics programs can also be modified, beginning with the regular mathematics education curriculum guide for the elementary school, which includes objectives and content topics. Content is selected for a total school mathematics program for the learning-disabled child. The premise that the child, regardless of age or grade placement, should be learning material that he or she is capable of learning is used as one guideline for content selection. Levels should be identified that contain behavioral objectives, lesson suggestions, sample assessment items, and commercially available materials, all to serve as a model. Decisions are made concerning the mathematical concepts and skills necessary for the child, the learning potential of the child, and the scope and sequence of the resulting content.

While such a selection and development of a curriculum is a large task, a particular school can begin with a smaller task of taking a scope and sequence chart from projects developed for learners with handicaps in learning and achievement and comparing this content list with the program in use.

STRANDS OF CONTENT

The Delphi approach and other attempts to develop new programs or modify existing programs strongly point toward a continuous progress program in mathematics. Strands of content should include multiple teaching and learning opportunities in the following areas: (1) numeration and whole numbers, (2) fractions and decimals, (3) geometry, (4) measurement, and (5) topics such as ratio, proportion, and percents. Basic and unifying ideas would include patterns, sets, operations, and relationships, all with the underlying theme of problem solving.

Geometry

Geometry involves position or location in space and serves to describe the world in which we live. The different ways in which we think about *where* something is located and the relationships that exist reflect the different kinds of geometries that have been developed. Those geometries that are directly related to a learner's experiences include (1) topology, in which the greatest amount of freedom is allowed in that few restrictions on objects exist in regard to size and shape, (2) plane geometry, which involves figures (shapes) and their relationships on a flat surface, and (3) measurement.

Numeration and Whole Numbers

The set of whole numbers, including the operations and properties involved with this set, constitutes a major component of middle/junior high school mathematics. Readiness is an essential consideration for beginning work with numbers. It seems evident that this area should not be the first strand that learners encounter in mathematics. Experiences with objects involved in sets, patterns, and geometry should serve as prerequisites for work with numbers. The concepts of cardinality, order, and place value should precede the actual operations, and the learners should be exposed to a wide variety of input-output modalities and simple problem-solving strategies before the operation algorithms are presented in symbol form.

The operations should be developed completely in carefully sequenced steps, that is, addition should be developed from putting objects together to develop basic facts; then examples involving no regrouping can be followed by increasingly more complex regrouping problems. Subtraction should be related to (defined in terms of) addition and developed in a similar manner. Multiplication should be defined in terms of repeated addition and sequenced properly. Division relates both to multiplication and subtraction.

It is in this strand of whole numbers where disabilities begin to appear as a result of mathematical gaps in the development background of the learners. The foundation must be solid or faulty algorithms may result. We often create the gaps by working with symbols before a learner is ready for the graphic symbolic stage, and if we *start* at this point the learner has no "fall-back" position.

Fractions

The fraction strand serves to extend the learner's concept of numbers to include numbers that represent parts of a group or parts of a whole thing. These new numbers form a large part of daily existence, from talking about distance to measuring quantities involved in cooking; and these new numbers cause the most difficulty for learners.

Basic to any real understanding of fractions is an awareness of parts, wholes, and the part-whole relationship; consequently, much attention must be given to laying the proper foundation.

Measurement

To provide the learner with a complete program of mathematics relevant to his or her needs, we must include appropriate measurement concepts. The social utility and necessity of the content is readily recognized by anyone faced with the management of independent living.

The measurement material that is usually covered as enrichment is too necessary and important to be left to chance. The concepts of measurement can be easily confused if not dealt with systematically and logically; and they must be reinforced by using a number of different related activities. Using different modes of instruction is important in this strand as well as in the other strands.

Other Topics

In addition to the more common elements of mathematics, the learners should be exposed to such topics as ratio, proportion, and percents in a manner that will enhance success and provide background to continue the study of mathematics. Elements of statistics and probability may also be included.

A PROPOSED SCOPE AND SEQUENCE FOR THE UPPER GRADES

In defining the mathematics content for the upper grades, outlines of topics for each of the content strands will be presented. Each outline will encompass the scope and sequence of the particular strand but the "horizontal" connection to other strands must be made by individuals designing their own programs. The development, for example, of numbers in decimal form (listed under Fractions) should not follow the complete development of fractions but rather be in conjunction with this topic.

Numeration and Numeration Systems

I. Prerequisite knowledge
   A. Counting
   B. One-to-one correspondence
   C. Reading and writing number values
   D. Cardinal and ordinal numbers
   E. Comparing numbers (=, >, <)
   F. Place value through 100

II. Place value
   A. Expanded notation
      1. Multiples of 10
      2. Using exponents
   B. Rounding to nearest 10, 100, 1,000
   C. Exponents
   D. Scientific notation

III. Fraction notation
   A. Numerator/denominator
   B. Mixed numerals
   C. Comparing fractions
   D. Ordering fractions

IV. Decimal notation
   A. Place value
   B. Rounding to the nearest 10th, 100th, 1,000th
   C. Scientific notation

V. Fractions to decimals and decimals to fractions

Whole Numbers

I. Prerequisite knowledge
   A. Addition
      1. A joining process
      2. Horizontal and vertical forms
   B. The addition algorithm
      1. Rounding addends and estimating sums
      2. Place value and regrouping

- C Subtraction
  1. A separation process
  2. Horizonal and vertical forms
- D. The subtraction algorithm
  1. Rounding minuend and subtrahend and estimating differences
  2. Place value and regrouping
- E. Multiplication
  1. Multiplication as repeated addition
  2. Arrays for multiplication
  3. Basic multiplication facts

II. Whole number addition
- A. Addition algorithm
- B. Properties of addition
- C. Alternative addition algorithms

III. Whole number subtraction
- A. Relationship of addition to subtraction
- B. Subtraction algorithm
- C. Properties of subtraction
- D. Alternative subtraction algorithms

IV. Whole number multiplication
- A. Multiplication as repeated addition
- B. Multiples
- C. Factors and products
  1. Prime factors
  2. Composite numbers
- D. Multiplication algorithm
  1. Rounding multiplicand and multiplier and estimating product
  2. Place value, regrouping, and partial products

V. Whole number division
- A. Division as repeated subtraction
- B. Division as the inverse of multiplication
  1. Products, factors, and quotients
  2. Division with no remainder, division basic facts
- C. Division involving quotient and remainder
  1. Remainder form
  2. Mixed number form
- D. Division algorithm
  1. Rounding dividend, division, and estimating quotient
  2. Place value, trial divisors, and multiples

Nonmetric Geometry

I. Prerequisite knowledge
   A. Patterns involving size and shape
   B. Parts of plane figures

II. Polygons and their characteristics
   A. Paths
   B. Sides and vertices

III. Lines, line segments, rays, and angles
   A. Parallel lines
   B. Perpendicular lines
   C. Types of angles

IV. Symmetry

V. Squares and rectangles
   A. Axes of symmetry
   B. Diagonals

VI. Family of quadrilaterals

VII. Triangles
   A. Altitude
   B. Angles

VIII. Circles
   A. Radius and diameters
   B. Circumference

IX. Similarity

X. Congruence

Fractions

I. Prerequisite knowledge
   A. Numerator/denominator format
   B. Fraction as a part of a unit (whole)
   C. Fraction as a portion of a group of objects
   D. Addition of fractions with like denominators

II. Equivalent fractions
   A. Fractional parts of units
   B. Fractional parts of groups

III. Reducing and ordering fractions

IV. Fractional parts of whole numbers (introduction to multiplication)

V. Proper and improper fractions and mixed numbers

VI. Additions of fractions
   A. Estimating size of answer in reference to given example as a prerequisite to addition of this example
   B. Like denominators
   C. Unlike denominators
   D. Addition of fractions

VII. Subtraction of fractions
   A. Estimating size of answer in reference to given example as a prerequisite to subtraction of this example
   B. Like denominators
   C. Unlike denominators
   D. Subtraction algorithm

VIII. Multiplication of fractions
   A. Estimating size of answer in reference to given example as a prerequisite to multiplication of this example
   B. Multiplication algorithm

IX. Division of fractions
   A. Estimating size of answer in reference to given example as a prerequisite to division of this example
   B. Division algorithm

X. Decimals (as an extension of whole number operation)
   A. Decimal notation for money
   B. Ordering decimals
   C. Rounding decimals
   D. Addition of decimals
      1. Money format
      2. Addition algorithm for decimals

- E. Subtraction of decimals
- F. Multiplication of decimals
- G. Converting between fractions and decimals
- H. Division of decimals
    1. Decimal number by whole number ($\neq 0$)
    2. Whole number by decimal numbers

Measurement

- I. Prerequisite knowledge
    - A. Arbitrary units of measure
    - B. Time
    - C. Money
    - D. Temperature
    - E. Relative size

- II. Linear measure for both English and metric systems
    - A. Segments
        1. Estimation
        2. Verification, the measuring process
        3. Rounding
    - B. Angles
        1. Estimation
        2. Verification
    - C. Perimeter and circumference

- III. Mass in both English and metric systems
    - A. Estimation
    - B. The measuring process

- IV. Area
    - A. Polygons and their regions
    - B. Circular regions

- V. Volume and capacity
    - A. Solids
    - B. Containers

- VI. Congruent relationships
    - A. One-to-one correspondence
    - B. Corresponding parts

Ratio, Proportion, and Percent

I. Ratio
  A. Ratio as a quotient
  B. Ratio as an ordered pair

II. Proportion
  A. Proportion relations
  B. Determining proportions
  C. Solution sets for proportions
  D. Finding missing components of a proportion relation

III. Percent
  A. Ratio to proportions with one denominator equal to 100
  B. Percents as hundredths
  C. Operations with percent
    1. Adding and subtracting percents
    2. Percent of numbers by multiplication
    3. Parts of wholes as percents
    4. Finding the whole from a given percent
  D. Applications of percent
    1. Percent increase and decrease
    2. Percent used in selling
    3. Percent used in interest
    4. Percent used in borrowing

DETERMINING METHOD

Another path in our task-analysis procedure certainly deals with the abilities and disabilities of the learners and we must consider this path as the method variable comes into the situation. By definition, a learning disability means a disorder in one or more of the basic psychological processes involved in understanding or in using language, spoken or written, which may manifest itself in an imperfect ability to listen, think, speak, read, write, spell, or do mathematical calculations. Since the study of mathematics is essentially the study of a language of symbols, the definition becomes relevant to this discussion. In attempting to answer the original question of what concepts and skills various learners possess we must also ask the question "How was the material presented?" Other questions are: What strategies and methods were used? Did the teaching style of the instructor match the learning styles of the learners? Was a developmental approach used?

To get us thinking developmentally (and later diagnostically) we will branch to an example of planning for developmental teaching that will indicate decision making and strategies—two prerequisites for determining method. First of all we need a definition.

DEVELOPMENTAL TEACHING OF MATHEMATICS

Developmental teaching is an instructional process that involves selecting and using strategies and methods that take into account the developing characteristics of learners. In developmental teaching of mathematics the added dimension of a logical sequence of content is included in the decision-making process. The instructional procedures include the *input* behaviors of the teacher and the *output* behaviors of the learners in various interactive combinations. For example, the most common interactive is for an instructor to select or develop a worksheet and present this to the learners as a group and the learners respond with written answers to the questions or exercises on the worksheet. The input information is symbolic (in written form) as is the output information.

Delimiting the instructional procedures only to this interactive reduces the opportunity of communicating with all of the learners at one time. Using alternative approaches is really not a new idea since many experienced teachers alter instructional procedures for different groups of learners and in different situations. This constitutes the first step in developmental teaching. Other steps require that we examine the characteristics of learners and the structure of content as we attempt to develop an effective match among learners, content, strategies, and methods.

If this sounds like a simplistic idea, and it is, then why aren't we teaching developmentally? In classes concerning methods in the teaching of elementary school mathematics some type of developmental approach is often presented; however, since the development is presented together, all at one time, and the prospective, or inservice, teacher does not see such a development in the textbooks, a developmental approach is viewed as something that exists in theory but not in practice. Many textbooks contain portions of a developmental approach, but this may be spread over two or three grade levels and thus in different books. In addition, past experience and formal course work have served to imply that each instructor of mathematics should seek closure, that is, each instructor should complete the development of a given concept or skill. If this happens with some concepts or skills, the time needed for development is lost by compacting one, two, or even three years of the developmental work into a shorter period of time. Surely, kindergarten and the first grades are developmental periods and closure should not be attempted with many of the concepts.

Teaching mathematics developmentally is a team effort with each member of the team providing just a few steps of the developmental sequence that begins in kindergarten and possibly concludes in either grade 7 or 8 in regard to general mathematics for the average and above-average learners. Concepts and skills are introduced and expanded in a spiral manner during these eight or nine years to develop a high level of understanding. Learning-disabled individuals will take a longer period of time to achieve a level of understanding.

An illustration of the developmental approach will be presented below using either statements or questions as the major divisions. In addition, examples will be provided at key points in the development and some suggested answers will be given to the questions. Bloom's (1968) taxonomy will be cited.

I. Determine criterion behavior(s) (desired learner output) with respect to given concept or skill.
Example: Concept: Addition of three three-digit addends with two regroupings.

$$\begin{array}{r} 284 \\ 176 \\ +\ 315 \\ \hline \end{array}$$

A. With respect to a content taxonomy (or content structure) where does this appear? [Bloom's taxonomy: application level]

B. What general behaviors will be expected of the learners?
1. Rounding each addend to nearest hundred
2. Estimating the sum
3. Speculating whether the estimated answer will be greater than or less than the actual answer (for students of "average" ability and above).

$$\begin{array}{r} 284 \rightarrow 300 \\ 176 \rightarrow 200 \\ 315 \rightarrow 300 \\ \hline 800 \end{array}$$

[Since two addends were rounded up, the estimated sum should be larger.]

C. What specific behaviors (concepts or skills, in this case binary and regrouping operations) are involved in the given situation?
1. Recognizing that an addition problem exists.
2. Knowing that the standard algorithm involves working from right to left.
3. Adding addends in ones column.
   (a) $4 + 6 = 10$
   (b) Maintaining the 10 in memory and adding 5 to it to obtain 15 ($10 + 5 = 15$).
   (c) Other combinations (basic facts) are also possible, for example: $5 + 6 = 11$; $11 + 4 = 15$.
   (d) A set of basic facts is implied: $\begin{array}{r} 4 \\ +5 \\ \hline \end{array}$ $\begin{array}{r} 4 \\ +6 \\ \hline \end{array}$ $\begin{array}{r} 5 \\ +6 \\ \hline \end{array}$ $\begin{array}{r} 5 \\ +4 \\ \hline \end{array}$ $\begin{array}{r} 6 \\ +4 \\ \hline \end{array}$ $\begin{array}{r} 6 \\ +5 \\ \hline \end{array}$
4. Recognizing that the sum of the addends in the ones column is one group of tens and five groups of ones, place the 5 of 15 in the ones position under the line and carry or regroup the one group of tens above the tens column—in this case above the 8.

$$\begin{array}{r} 1 \\ 284 \\ 176 \\ +\ 315 \\ \hline 5 \end{array}$$

5. Adding addends in tens column.
   (a) $1 + 8 = 9$
   (b) Maintaining 9 in memory and adding 7 to it to obtain 16. ($9 + 7 = 16$)

(c) Maintaining 16 in memory and adding 1 to it to obtain 17. $(16 + 1 = 17)$

(d) A set of basic facts is implied: $\begin{array}{r}1\\+1\\\hline\end{array}$ $\begin{array}{r}1\\+7\\\hline\end{array}$ $\begin{array}{r}1\\+8\\\hline\end{array}$

$\begin{array}{r}1\\+9\\\hline\end{array}$ $\begin{array}{r}1\\+15\\\hline\end{array}$ $\begin{array}{r}1\\+16\\\hline\end{array}$ $\begin{array}{r}2\\+7\\\hline\end{array}$ $\begin{array}{r}2\\+8\\\hline\end{array}$ $\begin{array}{r}8\\+8\\\hline\end{array}$ $\begin{array}{r}8\\+9\\\hline\end{array}$ . . .

6. Recognizing that the sum of the addends in the tens column is one group of hundreds and seven groups of tens, place the 7 of 17 in the tens position under the line to the left of the 5 in the ones position and carry or regroup the one hundred above the hundreds column—above the 2 in this case.

$$\begin{array}{r}11\\284\\176\\+\ 315\\\hline 75\end{array}$$

7. Adding addends in hundreds column.
   (a) $1 + 2 = 3$
   (b) Maintaining 3 in memory and adding 1 to it to obtain 4. $(3 + 1 = 4)$
   (c) Maintaining 4 in memory and adding 3 to it to obtain 7. $(4 + 3 = 7)$
8. Recognizing that the sum of the addends in the hundreds column is seven groups of hundreds, place the 7 in the hundreds position under the line to the left of the 7 in the tens position.

$$\begin{array}{r}11\\284\\176\\+\ 315\\\hline 775\end{array}$$

[Estimated answer of 800 was indeed larger than actual sum.]

D. Was the condition an example, exercise, or problem?
   [A multiple-step example]

II. Determine the placement of this item in the mathematical sequence with respect to a given grade level.
   [Grades 4–5—Placement: Ages 9–10
   Grades 5–6—Placement: Ages 10–11
   (From: Connecticut State Board of Education–1981, *A Guide to Curriculum Development in Mathematics*)
   Answer would be *Grade 5 in first quarter of the year in first 40 lessons*.]

III. Determine if this placement "fits" the "normal" growth patterns of your learners in terms of affective, social, physical, cognitive, and psychomotor development.

IV. If adjustments are to be made in terms of disagreements, what will be the guiding criteria?

V. Determine the easiest symbolic fact that is a part of the given situation. $\begin{array}{r}2\\+1\\\hline\end{array}$

A. With respect to the content taxonomy where does this appear?
   [Bloom's taxonomy: knowledge level]

B. What general behaviors will be expected of the learners?
   [Instant recall.]

VI. Determine the placement of this item with respect to a given age or grade level.
   [Grades 2–3—Placement: Ages 7–8
   (From: Connecticut State Board of Education–1981, *A Guide to Curriculum Development in Mathematics)*
   Answer would be *Grade 1 in second quarter of the year*.]

VII. Determine the developmental procedures necessary to teach 2 + 1 = 3.

A. Prerequisite knowledge (concepts) and skills.

1. Review of Addition and Subtraction Development

a. Steps

(1) Addition

Teacher Input	*Learner Output*
(a) Represent addends with objects.	
(b) Ask learner to write or find the numeral that tells how many in each set. (If for some reason or indication learners hesitate or cannot do what is requested review 1-to-1, counting, naming before going on.)	(b) Learner labels or writes numerals to identify numbers in addends.
(c) Ask the learner to move the blocks (for both addends) together—join the two sets of blocks—in a group to the right of where they were.	(c) Learner moves and groups blocks.
(d) Ask the learner to write or find the numeral that tells how many in the new set.	(d) Learner labels or writes numeral for the new set.
(e) Do these for: 1 + 1 2 + 1 3 + 1 4 + 1 1 + 2 2 + 2 3 + 2 1 + 3 2 + 3 1 + 4	
(f) Select pictures of sets of objects and present to the learners.	
(g) Ask learner to write or find the numeral that tells how many in each set.	(g) Learner labels or writes numerals to identify numbers for addends.
(h) Ask the learner to either trade the two pictures of the addends for one picture that shows all of the elements together and place this to the right of where the other pictures were *or* draw arrows to show the joining of the pictured objects together at the right of where they were.	(h) Learner demonstrates grouping of addends.
(i) Ask the learner to write or find the numeral that tells how many in the new set.	(i) Learner labels or writes numeral for the new set.
(j) Do these for: 1 + 1 2 + 1 3 + 1 4 + 1 1 + 2 2 + 2 3 + 2 1 + 3 2 + 3 1 + 4	
(k) Repeat the combinations alternating objects and pictures and develop each	

math fact in sentence form, e.g.,
$3 + 2 = 5$

[This development would occur before the beginning of grade 2.
Introduce "0" as an addend.
Either develop sums to 9 or go to subtraction.]

(2) Subtraction

Teacher Input	*Learner Output*
(a) Represent minuend with objects.	
(b) Ask learner to write or find the numeral that tells how many objects are given.	(b) Learner labels or writes numeral to identify numbers in minuend.
(c) Ask the learner to move (remove) some of the blocks to the right of the first set.	(c) Learner moves and groups blocks.
(d) Ask the learner to write or find the numeral that tells how many in the new set.	(d) Learner labels or writes numeral for the new set.
(e) Ask the learner to move the objects that remained in the first set to the right of the new set.	(e) Learner moves and groups blocks.
(f) Ask the learner to write or find the numeral that tells how many in the new set.	(f) Learner labels or writes numeral for the new set.

(g) Do these for:
$5 - 4$ $4 - 3$ $3 - 2$ $2 - 1$
$5 - 3$ $4 - 2$ $3 - 1$
$5 - 2$ $4 - 1$
$5 - 1$

(h–m) Repeat the process with pictures.

(n) Repeat the combinations alternating objects and pictures and develop each math fact in sentence form, e.g., $5 - 3 = 2$.
Introduce "0" as a subtrahend.

VIII. Use task-analysis model to determine a sequence (or possible sequences) that will go from:

$$\begin{array}{r} 2 \\ \underline{+\,1} \end{array} \quad \text{to} \quad \begin{array}{r} 284 \\ 178 \\ \underline{+\,351} \end{array}$$

Developmental teaching must involve (1) building on what the learners already have acquired, (2) using interactive modes that fit the learners (see Chapter 10), and (3) "filling in" any gaps that are noticed, for example, using objects to demonstrate a concept. "Building" the concept of adding three three-digit numbers with two regroupings might be developed at grade 4 or 5 in the following manner.

Teacher Input	*Learner Output*
State: "Today we are going to discuss addition that involves regrouping."	
Write: 7 + 2 State: "Do you remember how we did this using objects?"	A learner explains that "seven" blocks are moved together with "two" blocks to get "nine" blocks or "7 + 2 = 9."
[Note: As examples are built only the significant portion of the problem is exposed.]	
Write: 9	
Write: 40 + 30 State: "How do we do this example?"	A learner explains the movement of "4" groups of ten and "3" groups of ten to obtain "7" groups of ten or "40 + 30 = 70."
Write: 70	
State: "Let's put these examples together."	
Write: 40 + 30 = 70 and 7 + 2 = 9. They become 40 + 7 + 30 + 2	
State: "If we still add the number values, what do we get?"	"70 and 9"
Write: 70 9 or 79	
[Note: We could demonstrate with objects (e.g., Dienes blocks) if we deem it necessary.]	
State and demonstrate: "Let's combine the examples." 47 + 32 "One way that we might obtain the answer is . . . 7 plus 2 is 9."	
47 + 32 Write the 9 — 9 — and 40 plus 30 is 70. Write the 70 — 70 — Now add the 70 and 9 to get 79. Write the 79 — 79	
"This is the long way to complete the example and we want to write our answer	

directly, but this will help us when we regroup. You do the example as we usually do it."

$$\begin{array}{r} 47 \\ +\ 32 \\ \hline 79 \end{array}$$

[Note: The entire development looks like this:

(1)
$$\begin{array}{r} 40 \\ +\ 30 \\ \hline 70 \end{array} \quad \begin{array}{r} 7 \\ +\ 2 \\ \hline 9 \end{array}$$

(2)
$$\begin{array}{r} 40 + 7 \\ +\ 30 + 2 \\ \hline 70 + 9 \\ \text{or} \\ 79 \end{array}$$

(3)
$$\begin{array}{r} 47 \\ +\ 32 \\ \hline 9 \\ +\ 70 \\ \hline 79 \end{array}$$

(4)
$$\begin{array}{r} 47 \\ +\ 32 \\ \hline 79 \end{array}$$

State:

"Now let's do this procedure with

$$\begin{array}{r} 59 \\ +\ 83 \end{array}$$"

Guide the learner through the first three steps again.

$$\begin{array}{r} 50 \\ +\ 80 \\ \hline 130 \end{array} \quad \begin{array}{r} 9 \\ +\ 3 \\ \hline 12 \end{array} \qquad \begin{array}{r} 50 + 9 \\ +\ 80 + 3 \\ \hline 130 + 12 \\ \text{or} \\ 142 \end{array} \qquad \begin{array}{r} 59 \\ +\ 83 \\ \hline 12 \\ 130 \\ \hline 142 \end{array}$$

State:

"As the examples become longer and we obtain more than ten of any one unit (10s, 100s, etc.), we need to regroup to get the example in the form that we can list the answer immediately under the example. We regroup or carry."

Cover all of the example except what is in the unit's place.

$$\begin{array}{r} \square 9 \\ +\ \square 3 \\ \hline \end{array}$$

State:

"Nine (9) plus three (3) is equal to what number?" — "Twelve."

"The twelve (12) can be written as how many tens and how many ones? — "One ten and two ones."

Write the 2 (ones) under the line and under the 3. Remove the cover and "carry" the 1 to the tens column above the 5. Ask the learners to do the same.

$$\begin{array}{r} 1 \\ 59 \\ +\ 83 \\ \hline 2 \end{array}$$

State:

"Now add the 1, 5, and 8 to determine how many groups of tens you now have." — "Fourteen tens."

"The 4 goes under the line, under the 8 and the 1 goes next to the 4 and to the left of it."

"Our answer is . . . ?" — "142."

The steps for 284 + 176 + 315 are:

$$\begin{array}{r} 200 \\ 100 \\ +\ 300 \\ \hline 600 \end{array} \qquad \begin{array}{r} 80 \\ 70 \\ +\ 10 \\ \hline 160 \end{array} \qquad \begin{array}{r} 4 \\ 6 \\ +\ 5 \\ \hline 15 \end{array} \qquad \begin{array}{r} 200 + \ \ 80 + 4 \\ 100 + \ \ 70 + 6 \\ +\ 300 + \ \ 10 + 5 \\ \hline 600 + 160 + 15 \end{array} \qquad \begin{array}{r} 284 \\ 176 \\ +\ 315 \\ \hline 15 \\ 160 \\ 600 \\ \hline 775 \end{array} \qquad \begin{array}{r} \overset{1\,1}{284} \\ 176 \\ +\ 315 \\ \hline 775 \end{array}$$

with the algorithm being built from what the learners already know. Instructor should verbalize, write, and demonstrate the process, provide the learners with a sample exercise, and work in this format until learners get the idea. Gradually, step 1 is eliminated, then step 2, and after considerable work, step 3. Since diagnostic-prescriptive teaching is the reverse of developmental thinking, we have a pattern for reversing our process. As can be seen from the example, early experiences with materials and alternative approaches are essential as we attempt to meet the needs of learners with disabilities.

With the development and understanding of the algorithm (rule) for addition the learner should be given the opportunities to use the newly acquired concepts and skills in (1) tables, charts, and graphs that depict real-life situations and (2) "word" examples, exercises, *and* problems of a realistic nature. The learner should realize that the tables, charts, and graphs are used to display information and indicate relationships.

> Example:
> The local film theater sold 930 tickets for the two showings of "E.T." on Sunday. For the one show on Monday 460 tickets were sold. On Tuesday, for two shows 850 tickets were sold. Draw a graph to show ticket sales for the three days.

Some learners will be able to construct the graph without help. Others will need a start, for example, with naming the dimensions of the graph. Still others will need a prepared grid on which to work and some will require assistance with both grid and questions as an instructor fills in the information. After the charts have been completed such questions as the following should be asked:

1. On what day were the most tickets sold?
2. How many showings of "E.T." occurred at the theater from Sunday through Tuesday?
3. How many tickets were sold altogether?

These are word exercises and word problems. It should be noted that for some precise questions distractions exist. For example, when you ask about the number of tickets, the number of shows serves as a distraction. Learners should have had prior experience with this in easier situations. The developmental sequence for such word usages begins with simple content concepts, simple sentences with simple subjects and objects with no distractions, and with a reading level two grades below the level in which the example or exercise is given. Under these conditions one circumstance can vary while the others remain constant. For example, we could go to a higher level concept while maintaining the same reading level with simple sentence structure and no distractors. The example given with question number 3 represents the addition of three three-digit numbers involving two regroupings, the same (or simple) subject and same (or simple) objects with a distractor. The development continues until multiple-step problems are created. Such situations can be described orally, in person or on tape, in written format, by the use of diagrams, or charts, or a combination of the input modalities in an attempt to involve as many learners as possible in the problem-solving situation.

Since it is impossible within this chapter to develop each algorithm in as complete a manner as the addition example, other examples will be shortened and notes will be supplied to suggest procedures with respect to method.

A Subtraction Algorithm

Since subtraction is a separation process, the learners should have previous experience with separating groups of objects; naming and labeling results; trading groups of tens for ones and vice versa and groups of tens for groups of hundreds and vice versa; performing all of the activities involving pictures; and reading, writing, and stating relationships involving place value.

Example:

$$\begin{array}{r} 502 \\ \underline{-\ 357} \end{array}$$

To "build" this example the learner should have experienced 400 − 300 using objects, pictures, and symbols; likewise for 90 − 50 and 12 − 7. The first step is to review these processes.

$$\begin{array}{r} 400 \\ \underline{-\ 300} \\ 100 \end{array} \qquad \begin{array}{r} 90 \\ \underline{-\ 50} \\ 40 \end{array} \qquad \begin{array}{r} 12 \\ \underline{-\ 7} \\ 5 \end{array} \qquad \begin{array}{r} 400 + 90 + 12 \\ \underline{-\ (300 + 50 + 7)} \\ 100 + 40 + 5 \\ 145 \end{array} \qquad \begin{array}{r} 490 + 12 \\ \underline{-\ (350 + 7)} \\ 140 + 5 \\ 145 \end{array}$$

$$\begin{array}{r|r|l} 4 & 9 & {}^{1}2 \\ -\ 3 & 5 & 7 \\ \hline 1 & 4 & 5 \end{array} \qquad \begin{array}{r} {}^{4\,9} \\ \not{5}\not{0}2 \\ \underline{-\ 357} \\ 145 \end{array}$$

Since the last two or three steps involve something new:

1. Demonstrate the process, using blocks or an abacus (something from the background of the learners), and ask the learners to do likewise.
2. Ask the learners to follow the process as you demonstrate it and tell what you are doing.
3. Ask the learners to write the answers and to compare the results from step to step.

Continued development and reinforcement activities might involve the following interactives.

Instructor	*Learner*
1. Construct (or build) with blocks or an abacus the total subtraction algorithm.	a. Copy what instructor does. b. Identify the process and product from pictures of various stages of development. c. Describe what instructor does. d. Write the numerals to represent the example being demonstrated.
2. Present pictures of the process and product stages of the algorithm.	a. Construct the example with blocks or on an abacus. b. Arrange the pictures of the developmental steps in order. c. Explain the process of subtraction. d. Write the steps for the subtraction process.
3. Explain the subtraction algorithm.	a. Demonstrate each step with blocks or on an abacus. b. Identify the steps from a series of pictures. c. Repeat the verbal description of the subtraction algorithm. d. Write each step of the process using symbols.
4. Provide a worksheet of the processes and product involved in the subtraction algorithm.	a. Construct the example with blocks or on an abacus. b. Find pictures to represent the process. c. Verbally describe the subtraction process. d. Write the answer to each example.

As each of the algorithms are "built" and suggested strategies and procedures are demonstrated, bear in mind that, with modifications, such strategies and procedures could be used with other sets of numbers and other algorithms.

A Multiplication Algorithm

After the basic facts of addition have been developed, reinforced, and understood, multiplication can be introduced as repeated addition. For example, 2 + 2 + 2 can be represented as three groups of 2 or 3 × 2, which can be constructed in array form as:

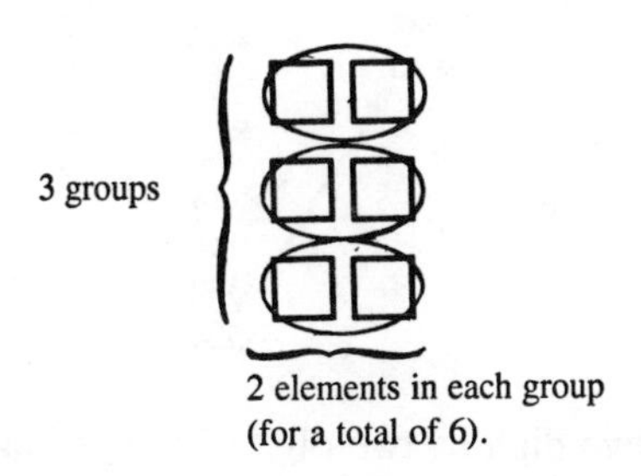

The extension of multiplication to the algorithm involves development of a good understanding of place value and regrouping. The development of a multiplication algorithm should follow the same developmental patterns as in the addition and subtraction algorithms. We will demonstrate one sequence for an example involving a three-digit multiplicand and a two-digit multiplier. With such a look at a total algorithm an instructor will be able to see where a learner could go wrong.

To develop the multiplication of 14 × 36 we might use the following strategy with the number and operational symbols.

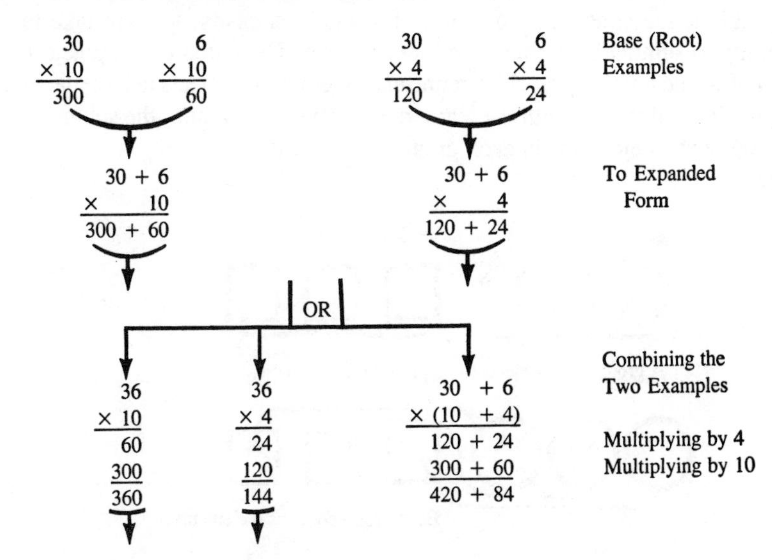

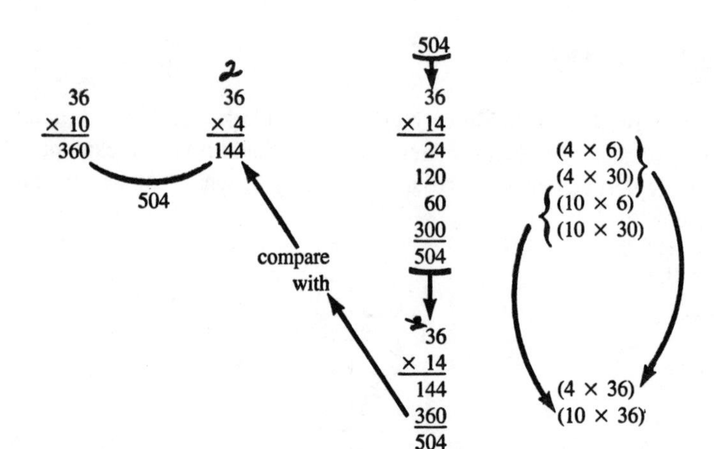

Development of this two-digit by two-digit with regrouping example involves putting together what the learners should already know and introducing the regrouping idea.

A Division Algorithm

The division concept was foreshadowed when learners sorted objects into categories, when they chose sides, when they "divided" straws, candy, etc., among the class members. The formal development should begin with objects to be grouped or sorted according to some precise rule. Relationships to multiplication and subtraction should be established while using pictures or objects.

Since this is the concluding operation for whole numbers, we will take this opportunity to summarize a total development idea. The initial development for division of whole numbers involves representing groups of objects to be separated into equivalent subgroups with the learners identifying the group, the subgroups, and the number of elements in each group.

Example:

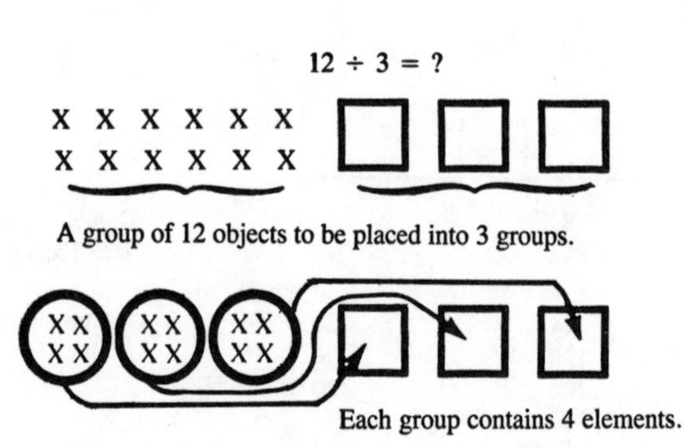

We should also connect such an example to its related multiplication example of $3 \times \underline{\ ?\ } = 12$. This example could be increased to $120 \div 30 = ?$ using a similar approach.

As the division algorithm is being developed, the learners should be able to (1) demonstrate the "dividing" (separating) process with objects (or on an abacus), (2) identify which set of displayed answers represents what is being developed, (3) state the step of the developed process, and (4) write a numerical expression to illustrate the process.

The separation process of filling groups leads into the "repeated subtraction" strategy for whole number division.

Example:

$54 \div 9$ (or $\frac{54}{9}$ or $9\overline{)54}$)

54		$9\overline{)54}$
− 9	one group of 9	− 9
45		45
− 9	one group of 9	− 9
36		36
− 9	one group of 9	− 9
27		27
− 9	one group of 9	− 9
18		18
− 9	one group of 9	− 9
9		9
− 9	one group of 9	− 9
0	6 groups of 9	0

Therefore there are 6 groups of 9 elements in 54 or

$$\frac{54}{9} = 6 \quad \left(\begin{array}{r} 6 \\ 9\overline{)54} \end{array}\right)$$

As this process is shortened to $\begin{array}{r} 6 \\ 9\overline{)54} \end{array}$ for the basic division fact we can build the large algorithm by putting such simple problems together. (Note: We should also introduce remainders as we extend the development.)

Example:

$3\overline{)36}$

The base examples are: $\begin{array}{r} 10 \\ 3\overline{)30} \end{array}$ and $\begin{array}{r} 2 \\ 3\overline{)6} \end{array}$

This becomes: $\begin{array}{r} 10 + 2 \\ 3\overline{)30 + 6} \end{array}$

This is extended to:

```
    2
   10   or 12
 3)36
   30
    6
    6
```

Finally, we get:

```
   12
 3)36
   3
    6
    6
```

Let's see how this process may eventually be extended to a larger example:

```
                         32)479

          30 + 2)400 + 70 + 9               in expanded form

                      14                    9) regrouping answer (quotient)
                   ⏞
1) There are ten    10 + 4                  5) There are four thirty-twos in (150
   thirties in (400 + 70)                      + 9).
              30 + 2)400 + 70 + 9

                     300 + 20               2) 10 (30+2) = 300+20
                     100 + 50 + 9           3) subtracting and bringing the 9
                     ⏟                         down
                        150 + 9             4) regrouping
                        120 + 8             6) 4(30+2) = 120 + 8
                         30 + 1             7) subtracting
                         ⏟
                           31               8) regrouping
```

answer is 14 with a remainder of 31

The next step in this process would be:

```
     4 } 14
    10
 32)479
    320
    159
    128
     31R
```

The final step is the standard algorithm.

```
     14
 32)479
    32
    159
    128
     31R
```

A Concluding Statement on Whole Numbers

Each algorithm for whole numbers has been presented using different steps of the developmental process and involving several interactives (combinations of teacher input information and student output information). Combinations of all of these steps and interactives would present a complete picture of whole number development.

As the examples become exercises and understandings result, problem solving can be introduced. Foreshadowing steps include rounding, estimating, and developing tentative hypotheses (making guesses). While we may use games and puzzles for motivational purposes, we need to create situations that resemble real world situations. From such simple activities as "Find the missing numbers," for example, $15 + \_\_\_ + \_\_\_ = 27$, we can develop skills that can later be used in such problem situations as "How much more will it cost in interest, if I pay for my car in 36 months instead of 24 months?" "What will be the value of the car in three years?" Seldom do we find only one answer possible in real life problems and we need to train learners to think in terms of best answers at a given time period. Reasonableness of solutions is also an issue, especially if our world is to be filled with such devices as microcomputers. What can we expect as a solution?

DEVELOPMENTAL TEACHING OF FRACTIONS

From an introduction to division involving whole numbers we obtain two strategies that outline numbers in fraction form: a measurement strategy and a partitioning strategy. Measurement division appears to be the easiest one for students to learn. This strategy involves determining into how many subsets a given set of objects can be divided (see Figure 7–3). For example, if I have ten pieces of candy (my given set of objects) and I decide to give two pieces of candy to my friends, how many friends will get two pieces of candy? I could serve five friends. A usual format for this example is $10/2 = ?$

The other strategy (illustrated in Figure 7–4) involves determining how many elements will be in a predetermined number of subsets. For example, if I have ten pieces of candy and want to share with five of my friends, how many pieces of candy will each get? Each of us will get two pieces of candy. Again, the example format is $10/5 = ?$ Either strategy foreshadows the introduction to fractions as we create denominators (the bottom or second number of a fraction). When real objects such as apples or blocks are used the denominator should be understood to represent how many parts a whole object or set of objects have been divided into (see Figure 7–5).

A learner should be able to group, name, find a label and write the symbol for the denominator whether objects or pictures of objects are being used. Paper

Figure 7–3 Measurement Strategy for Division

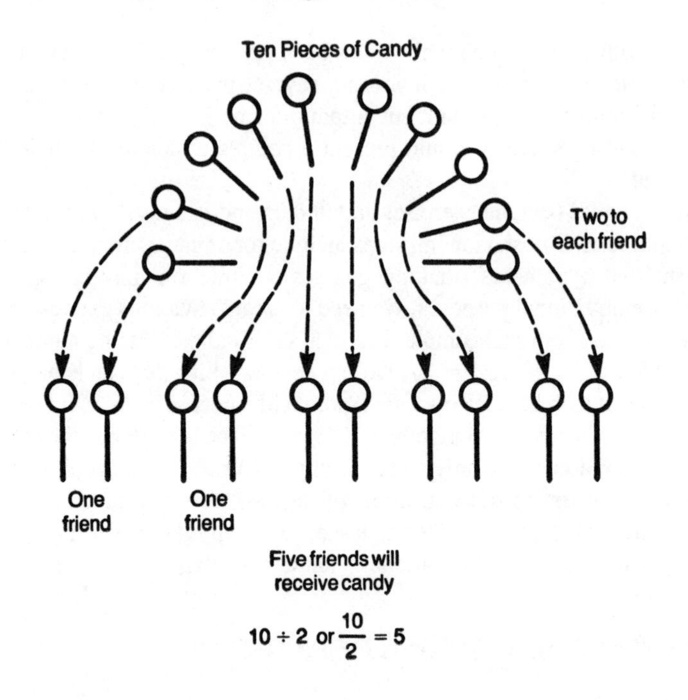

Figure 7–4 Partitioning Strategy for Division

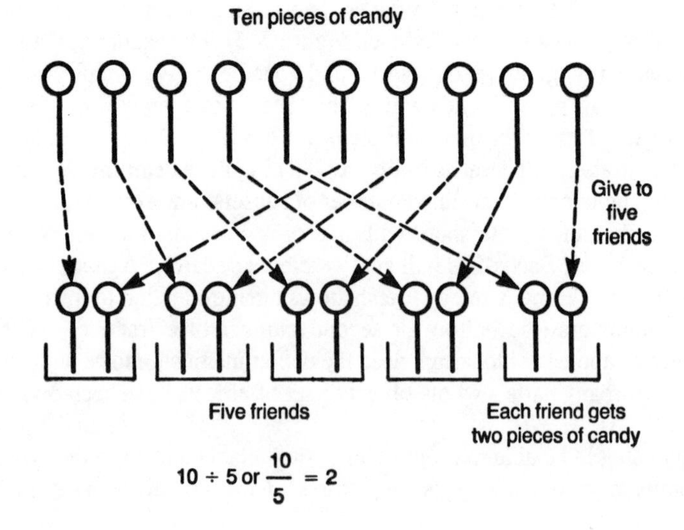

Figure 7–5 Division Strategies

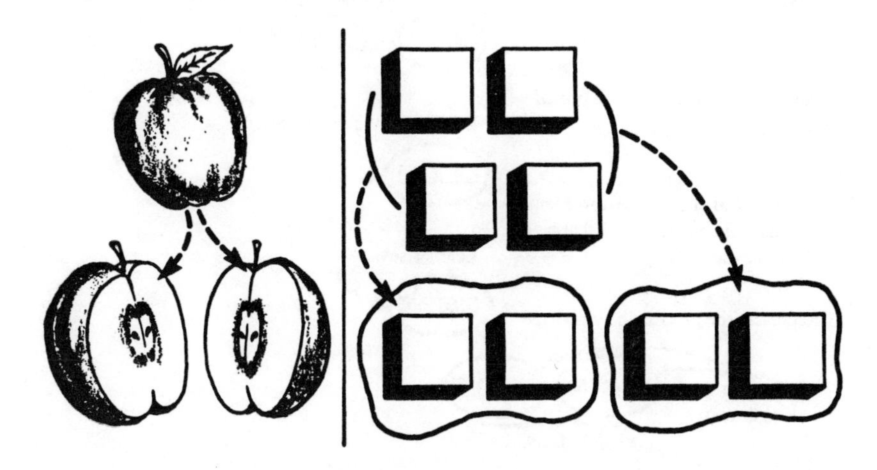

Two parts in both cases

circular or rectangular regions can be folded to create "parts of a whole." By shading sections of the regions the numerator (top of first number of a fraction) can be developed to mean "the number of parts being used." Thus, ¼ implies that we are using one of a possible four parts. Constructing and showing equivalent fractions becomes the next concept and this one leads into ordering fractions. Using objects, pictures, *and* symbols provides us with the best opportunity to *reach* the greatest number of learners. Models should be used as long as they serve to clarify the concept being developed. When there is a chance that the model might interfere with the concept, as in division of fractions, we should cease to use the model. No developmental steps should be omitted. Sample strategies will now be outlined.

Unit Fraction "Times" a Whole Number That Does Not Equal Zero

Example:

$$\frac{1}{2} \times 8$$

Teacher Input	*Learner Output*
Present eight blocks to the learners and ask them how many we have.	Learners name "8."
□ □ □ □ □ □ □ □	"8"
Ask a learner to separate the set of eight blocks into two subsets, each	

containing the same number of blocks.

Learner forms two equivalent subsets.

Now ask a learner to move one of the subsets.

Learner moves one subset.

Ask the learner to name how many objects there are in the set that he or she removed.

Learner says, "4."

Represent the process in symbols;

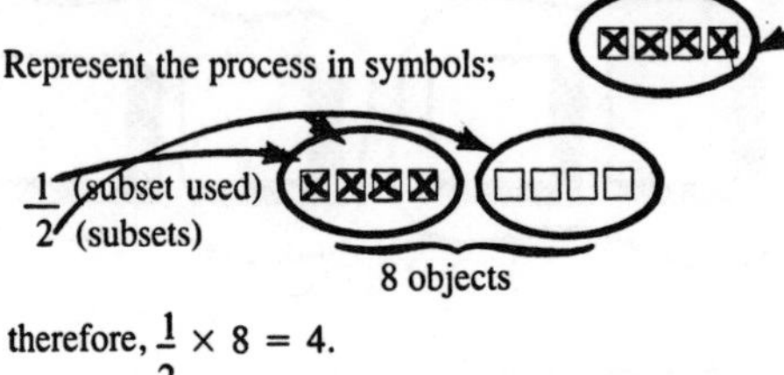

therefore, $\frac{1}{2} \times 8 = 4$.

Whole Number "Times" a Unit Fraction

Example:

$$4 \times \frac{1}{2}$$

Teacher Input	*Learner Output*
Present the learner with four circular regions with diameters drawn.	
Ask the learner to shade one of the two parts in each figure (½ of each figure). If paper regions are being used, have the learners cut out the parts.	
	Learner shades (or cuts out) one-half of each region.
Ask the learner to draw (or move) the one-half sections so that they fit together.	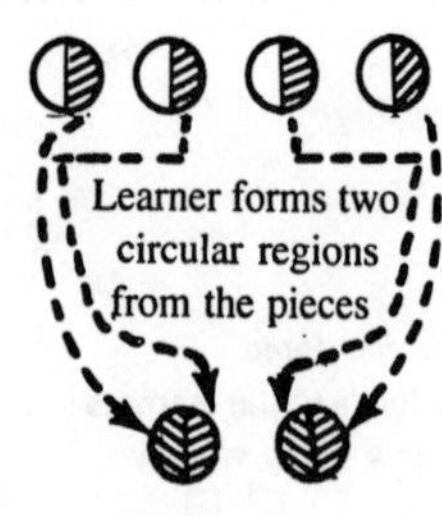
Ask the learner to tell what four one-halves are. Repeat the process and label each part.	"$4 \times \frac{1}{2} = 2$"

Unit Fraction "Times" a Unit Fraction

Example:

$$\frac{1}{2} \times \frac{1}{3}$$

Teacher Input	*Learner Output*
Ask the learners to take a strip of paper and fold it like a letter—into three parts of the same size (excess may be trimmed).	Learner folds strip of paper into thirds.

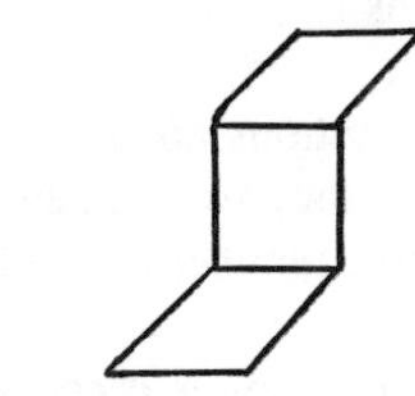

Ask the learner to shade one-third of the strip of paper.

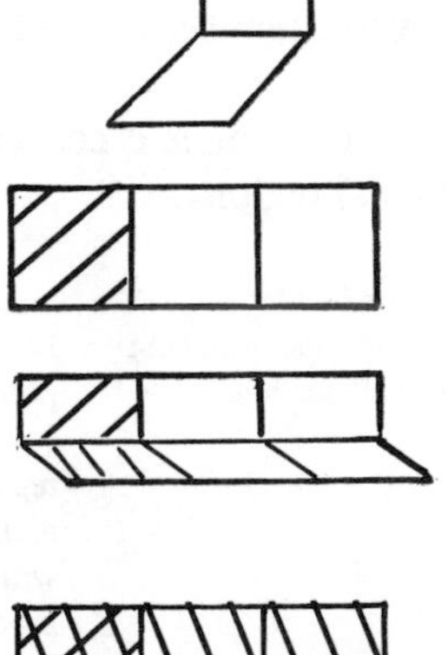

Ask the learner to now fold the same strip of paper into two parts of the same size, only this time fold widthwise.

Ask the learner to shade one-half of the strip of paper.

The strip of paper is now divided into six parts of the same size and one of the parts has been marked twice and represents one-half of one-third or one-sixth of the strip of paper.

$$\frac{1}{2} \times \frac{1}{3} = \frac{1}{6}$$

This developmental approach should be maintained until learners can generalize to the symbolic rules of operation (the algorithms). Addition and subtraction involves naming and labeling fractional parts, cutting out or shading these parts, moving the parts together (as in addition) or separating these parts (as in subtraction) to obtain the answer. Symbolic representation of the process constitutes a concluding step.

As we attempt to meet the needs of learning-disabled youth we need to sequence our developmental approach into small related steps. We must be able to give a complete verbal description of the process to obtain the desired answer—a verbal description that could stand alone without aids if it had to because of a dysfunction in visual perception. This signifies the degree of understanding that an instructor must have of a given process. Fractions and decimals are good examples in this domain. We must also be able to represent the process by development and manipulation of models without extensive verbalization if necessary. A good understanding of both of these extremes will enable an instructor to use combinations of these strategies more effectively.

Division of Fractions

There are some questions about the value of teaching division of fractions; however, to continue the developmental theme we will outline some possible interactives for a whole number divided by a fraction and a fraction divided by a whole number.
Example: $4 \div \frac{1}{2}$. This example is concerned with how many half sections are in four whole (complete) regions.

Teacher Input	*Learner Output*
Obtain four circular regions and cut each region into half sections.	Learners should be able to repeat your procedures; to identify pictures that show what you are demonstrating; name the number of regions, the fraction into which each region is divided, and the total number of parts; and write the numerical expression for the process: $4 \div \frac{1}{2} = 8$.
Use a square region. Fold it into half sections. Now, divide (by folding) one half section into four parts. Ask the learners to tell how large one of the small sections of the region is.	Learners should be able to repeat your procedure, to find diagrams and order them to explain the division process, state the appropriate division example, and write the numerical expression for the process: $\frac{1}{2} \div 4 = \frac{1}{8}$

When the final rule for division of fractions is presented each major step must be illustrated:

Example: $\frac{2}{3} \div \frac{5}{8}$

1) $$\frac{\frac{2}{3}}{\frac{5}{8}}$$

2) $$\frac{\frac{2}{3} \times \frac{5}{8}}{\frac{8}{5} \times \frac{8}{5}}$$

Use this format if you use multiplicative inverse strategy—that strategy that will make the denominator 1.

3) $$\frac{\frac{2}{3} \times \frac{8}{5}}{1} = \frac{2}{3} \times \frac{8}{5} = \frac{2 \times 8}{3 \times 5} = \frac{16}{15} \text{ or } 1\frac{1}{5}$$

4) 5) 6)

Again, after the basic algorithms have been developed, examples, exercises, and problems should be presented in word form. For example, you have half of an apple pie and five friends come to visit. How much will each of you get if you divide the one-half pie evenly among you? Is the portion large enough to suggest that you should offer the pieces of pie?

The base exercise is ½ ÷ 6 (five friends and you). This implies that each individual will receive one-twelfth of a pie. A "real" problem emerges. Should you offer the pie or is each piece so small that no one individual will receive an appropriate amount? Stating the last question might serve to make the exercise into a problem and, at the same time, create a realistic situation where subjective judgment is involved.

A LOOK AT DECIMALS

Background work with whole numbers operations, place value, and unit regions in fractions will serve us well as we extend the numbers to include decimals. Since numbers in decimal form are the numbers of calculators and computers, inclusion of this component is essential.

Beginning work with decimals should involve money, especially pennies, dimes, and dollars. Learners in grades 3 or 4 should learn the names of the various coins and the dollar, be able to count money, tell how much money is represented, and write the symbols (numeric) form of what value is represented. For example, given three pennies, four dimes, and one dollar, learners should be able to tell you that you have $1.43 and write the symbol for this amount. At this point of understanding the operations of addition and subtraction involving numbers in decimal form can be introduced with a review of the activities involving whole numbers. The important multiplication and division operations involve multiplication and division of a decimal number by a whole number, for example, $3.25 × 6 or $3\overline{)7.56}$. These can be developed using the same approaches as with whole numbers.

Some developmental activities that are necessary with decimals involve (1) writing fractions as decimals and vice versa, (2) comparing decimal values, (3) ordering decimals, and (4) extending the concept of place value. Since place value is necessary for an understanding of decimals, careful attention should be given to its development in the middle school. A counting board of one hundred regions (a ten by ten grid), graph paper of the same size, strips of ten regions, and single regions are excellent devices. The counting board may be defined as one unit and each small region as 0.01 (one hundredth) of the unit. Learners should be asked to manipulate the regions and color graph paper regions as they name, label, and identify tenths and hundredths. After an understanding occurs with these two place value concepts, an instructor can extend the development to thousands and beyond. With hundredths we have the background of percent as we define the one small square region of a large counting board as one percent (percent means hundredths)—one percent is equal to one hundredth (0.01).

DEVELOPMENTAL TEACHING OF GEOMETRY

Geometry is a natural content area for developmental teaching since we live in a three-dimensional world of height, width, and depth. A child's first movements are within these dimensions. Geometry is perhaps the only part of mathematics where most of the models are readily available for the learners to see. The concepts of size, shape, and color are easily demonstrated. Positions of objects can be described and descriptive words mean something. For example, objects or people have relative position and size if one is in front of the other or if one is tall and the other is short. Changing positions introduces paths, line segments, and lines—with all of their unique subsets.

Determining, through experimentation, which objects will roll along straight line paths allows learners to explore geometry. Tracing boxes and cans provide a firsthand experience in creating a square, a rectangle, or a circle. They do indeed have an inside and an outside as the figure is constructed. Attention to specific characteristics of three-dimensional figures leads to the study of shape.

All of this is prerequisite information for the study of shape if we want a developmental approach to occur. The procedures change from a descriptive approach to an analytical approach as we start giving more attention to detail.

Figures in a Plane

The development in geometry continues with an introduction to Euclidean geometry, which involves figures on a flat surface (in a plane). The square, rectangle, circle, and triangle are the first elements in this development. Learners should be able to identify, name, and draw these figures. Efforts should be made to relate these shapes to parts of real world objects. Efforts should also be made to show these objects in "nonstandard" positions so that learners will *not* develop a rigid view of each shape.

Figures to trace provide excellent opportunities for an instructor to relate simple closed paths to the two-dimensional figures. Each figure should be traced and the number of *sides* and *vertices* (corners) should be determined. If the development occurs in the middle school, the idea of *polygon* (a many-sided figure) should be introduced. This development should occur until each polygon and its characteristics are developed.

Other polygons, such as the rhombus, parallelogram, or other quadrilaterals, can be included next; however, we should not make the common mistake of losing contact with real objects. (Mathematics is a unique subject in that we can see a model, develop ideas about that model, change the ideas mentally in a logical manner, and then lose the concept of the original model. For example, we can imagine how to bend an object and develop a mental picture of the bent object, something we could not do with the object.) One way to avoid this is to reverse the ideas and build solids (three-dimensional figures) using each polygon as a base.

Sample Instructional Activity

Present a polygon pattern to each learner, a different one to each learner if you desire.

Ask the learners to trace the polygon pattern twice and cut each tracing out of the paper. Give the learners precut sides to be used to construct the figures.

Points and Line Segments

After the ideas of shape, relative size, and relative position have been developed to the extent described in the previous sections, an awareness of specific characteristics and relationships can be developed. The concepts of points and line segments are important because they serve to define that set of points that establishes the conditions for line measurement. Points can be established by pointing to the vertices or corners of the two- and three-dimensional figures. Each point is part of the figure or each point is *on* the figure. Other points can be located *inside* the figure or *outside* the same figure.

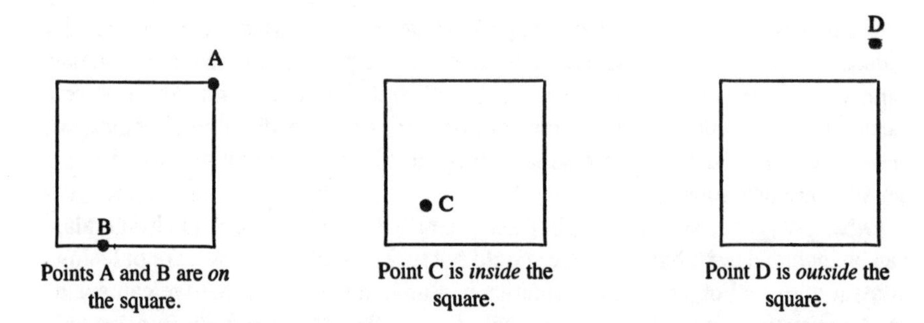

Points A and B are *on* the square.

Point C is *inside* the square.

Point D is *outside* the square.

The key idea in this development is that the point has *relative position*. This is its only characteristic, as a point has no definite size or shape. The light from a star in the sky can be considered a point and has relative position. A neutron can be a point and has relative position. This relative position comes into play when a portion of a straight line (or a straight path) is drawn from one point to another producing a *line segment*. Each side of a polygon is a line segment if the two points are included.

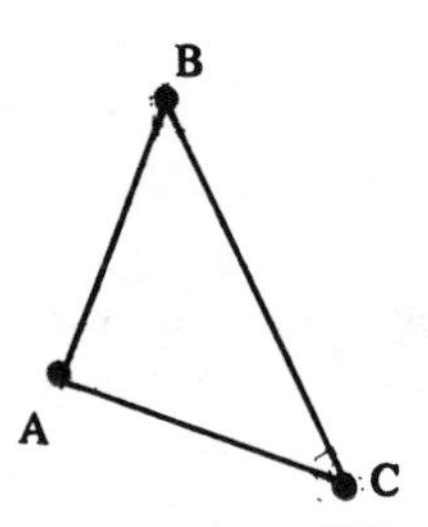

The triangle ABC has segments: $\overline{AB}$, $\overline{BC}$, and $\overline{AC}$.

The line segment has relative position and size. This size is named *length* or *distance* and involves measurement. Linear measure concepts are presented below.

The points at the ends of the line segments of the triangle serve to identify a corner or *vertex* of the triangle. Although it should be recognized that there are many points between the end points of a line segment, the points that identify a line segment and, in turn, identify the polygon they form, are the end points. It should also be noted that geometry is concerned with sets of points in certain locations (having relative position).

While it is desirable to maintain contact with the real world by identifying points, line segments between these points, and polygons formed by the line segments, our analytical approach has meaning by being easier to "see" if we draw a diagram (use the graphic symbolic modality) of the resulting figure.

Sometimes the attempt to create a real world situation results in an artificial situation for learners, thus causing more confusion than clarity. Evidence of this appears when learners fail to diagram or describe the "real world" situation accurately. A model should be used whenever possible; if we have no model we should describe and illustrate in detail and then go directly to a diagram.

We should also understand that the figures that are being used (square, rectangle, triangle) to develop precise characteristics are much more restrictive than what usually occurs with real world objects; for example, a simple paper clip. We take a rather large step from three-dimensional objects to two-dimensional objects in a restrictive manner. The characteristics become both unique and general—a vertex of a square is named by a point and a point can be identified anywhere on a paper clip. Another part-whole relationship is being developed.

It is at this point (sometimes a beginning point) that we can become too abstract for some learners. Perhaps we should question the value of such an analytical approach for *all* learners; however, if geometry is to be extended into such areas as art and carpentry we should examine its unique elements. With all of the previously described concepts and skills as prerequisite material, the chances for success are greatly increased and the chances of creating gaps are minimized.

Sample Instructional Activity

Present learners with models of cubes, rectangular solids, pyramids, prisms, spheres, cylinders, and cones.

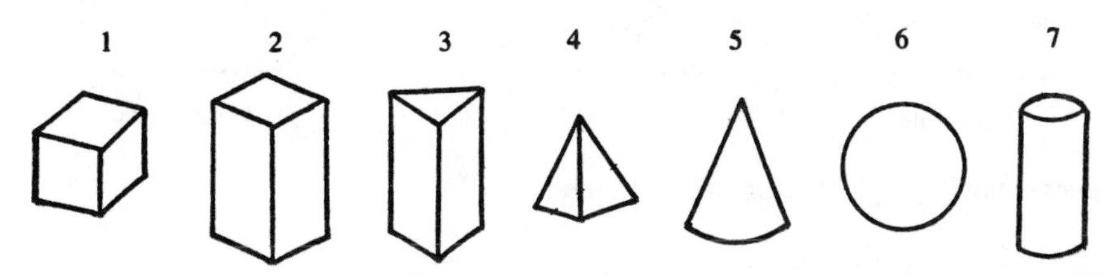

Ask them to determine the number of vertices, edges, and faces in each figure and record these numbers on a data sheet in the appropriate spaces (see Table 7–1).

Table 7–1 Data Sheet

	Three-Dimensional Figures		
Figure	*Number of Vertices* (Points)	*Number of Edges* (Line Segments)	*Number of Faces* (Planes)
1. (cube)			
2.			
3.			

Angles

Select a cube or rectangular solid, trace one side on a chalkboard, label the vertices A, B, C, and D respectively, and extend two adjacent sides.

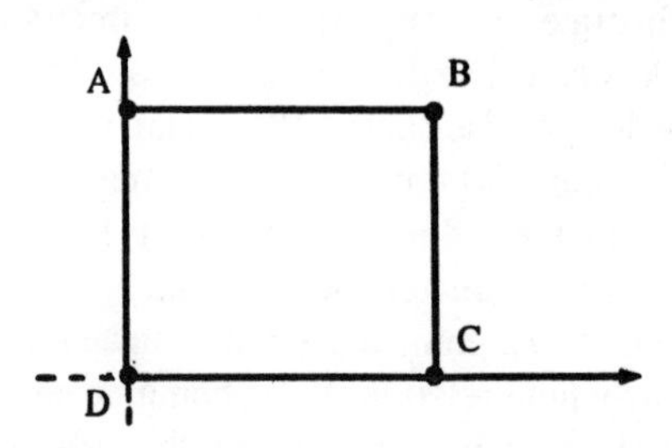

Explain that side AD and side DC are parts of two *lines* that cross (intersect) at point D. These sides are also part of *rays* that begin at D and extend in different directions. The two rays form an *angle* at point D. We are getting further removed from the real world and efforts should be made to identify objects that assume angle positions, for example: the "hands" of a clock. Angles in three types of positions should be considered.

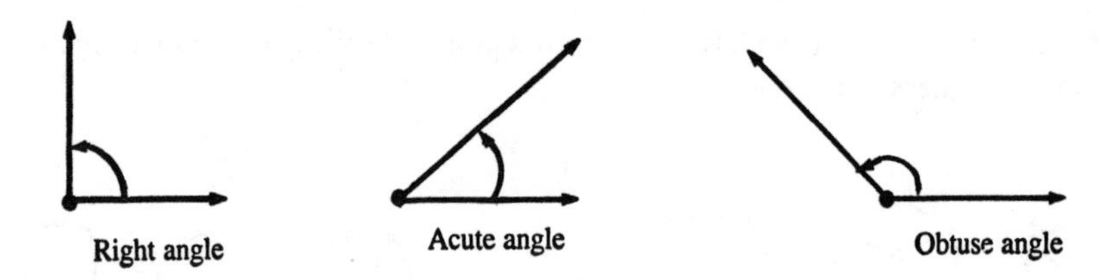

These should be pictured in known figures.

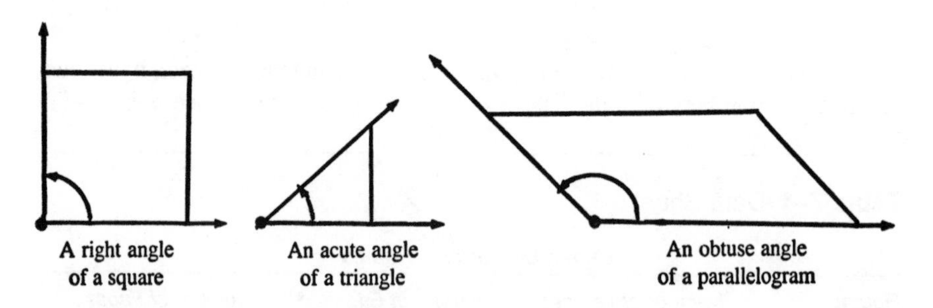

Learners should be asked to identify and construct these kinds of angles independently and within given polygons. Instructors should emphasize that the angles are the rays and *not* the opening.

With these background elements the learners are equipped to describe polygons in detail.

It is at this point that learners with handicaps in learning and achievement may begin to experience some difficulty and the concepts may become too difficult. If this happens, study should begin within the measurement strand instead of trying to develop more difficult concepts in this area.

Congruent Figures

Figures that have the same size and shape are said to be *congruent*. Sorting by these attributes provides learners with firsthand experience in identifying congruent regions (figures). Since the activity involved is one of comparing, it is easier than the analysis of polygons; therefore, the study of congruent figures is a viable alternative until learners are ready to continue working with vertices and sides. Work in this area also provides the opportunity to "fill the gaps" as many elements of geometry are reviewed in preparation for measurement. Chances for success increase with this activity.

Learners can gain experience in constructing congruent polygons by tracing the bases (or surfaces) of three-dimensional figures or patterns of polygon shapes.

Sample Instructional Activity

Present learners with sets of regions of the same shape but different sizes (at least three different sizes). Provide a worksheet with patterns drawn from the regions mixed on the page. Ask learners to identify the groups of congruent figures.

Activities such as this can lead into the topic of similarity where size is the variable.

Similarity

Similarity is a concept that is related to ratio and proportion. Its development should follow after that of ratio and proportion since, by definition, similar figures have the same shape and all sides are in the same proportions.

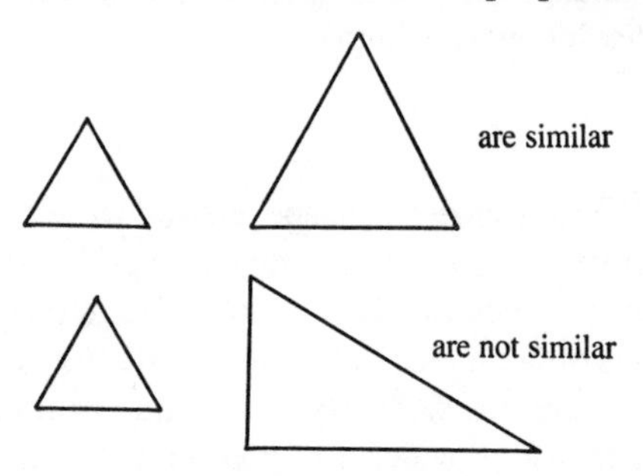

Identification is the desired output behavior here as accurate drawings are very difficult to obtain. Enlarging and reducing activities from photography involve the concept of similarity and similarity foreshadows the basics of projective geometry.

Symmetry

Symmetry relates to congruence in that we are looking for congruent parts within complete figures. The major question asked is how can a pattern of a given figure be folded so that the two parts will fit exactly together. For example:

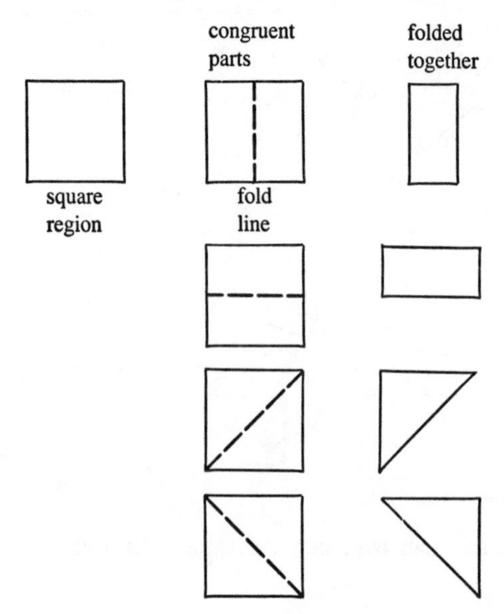

The square region has four different lines of symmetry. As the parts are folded together, they fit together with no overlap; therefore, they are congruent.

Sample Instructional Activity

Provide learners with patterns so that they may determine if lines of symmetry exist and if so how many such lines exist.

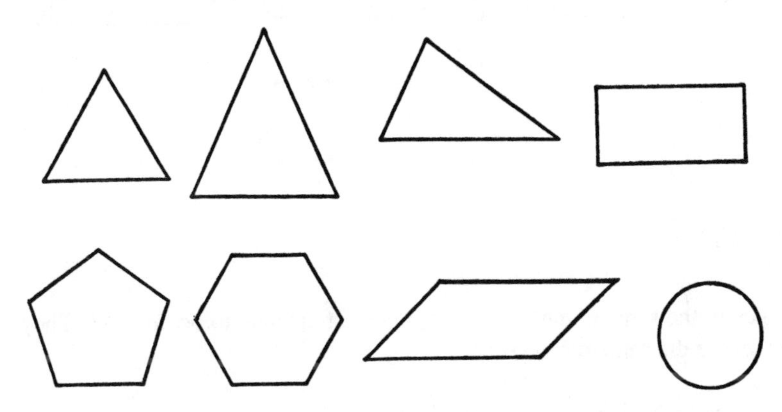

Symmetry is a significant element in design and a concept that learners can "experience;" therefore, it should be in the geometry curriculum.

Line Relationships

Much of the geometry to this point has been developed in an intuitive manner. To complete this development and to serve as a connecting link between this kind of geometric development and a more formal development, the study of the relationships between lines—intersecting, parallelism, and perpendicularity—should be undertaken. All the prerequisite work for these concepts has been developed in the previous sections; however, work above the intuitive level in this area should be delayed until the learners are approaching the formal stage of learning.

Intersecting Lines

As seen from the development of angles, each vertex of a polygon can be viewed as the (set of points in the) intersection of two straight lines; therefore, if lines meet, the result is an intersection through one point. The vertices of a square or rectangle involve unique intersections since right angles are formed with each intersection. If this occurs the lines are said to be *perpendicular*. Perpendicular lines form right angles.

Sample Instructional Activities

Present several polygons to the learners and have them select those polygons that have sides that are part of perpendicular lines.

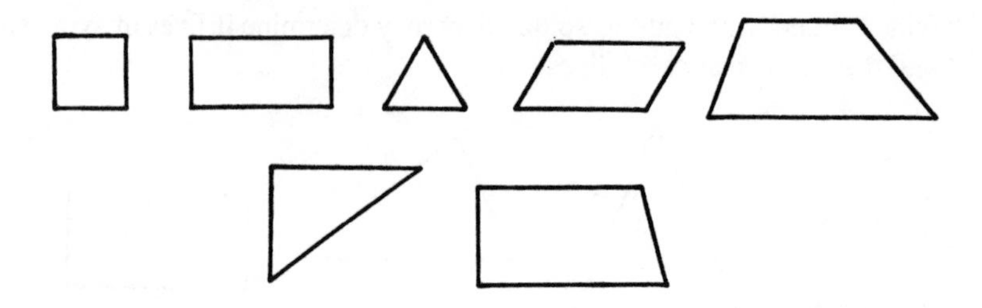

Parallel Lines

Lines in the same plane that do not intersect are said to be *parallel*. They are everywhere the same distance apart.

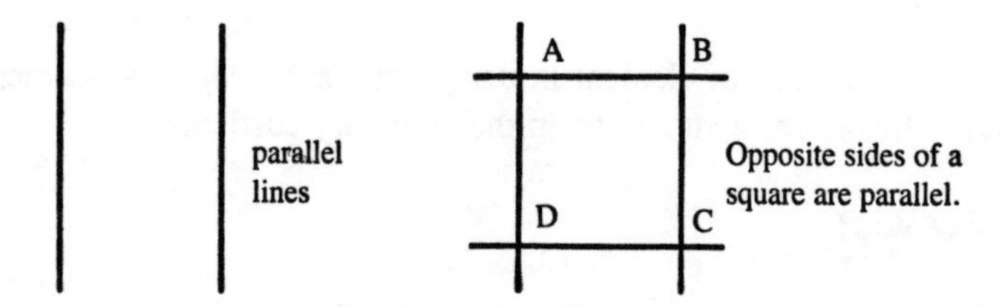

Sometimes it is necessary to extend the lines that contain the sides of polygons to see if they intersect. A trapezoid, for example, has one pair of parallel sides and one pair of sides that are not parallel because if extended, the lines containing these sides intersect.

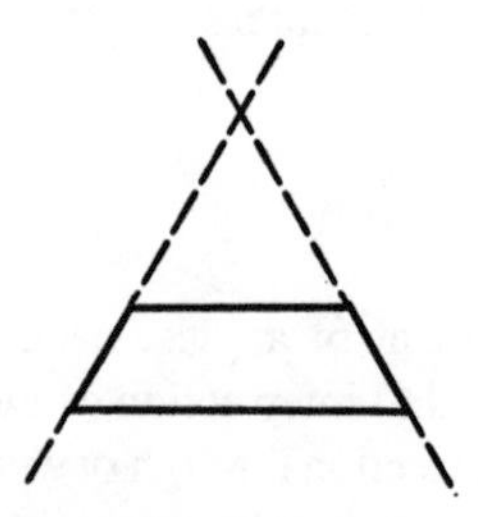

Learners should be asked to identify lines that are parallel or perpendicular or just intersecting. They should be able to construct lines of each type.

With this type of geometric background, a learner should be able to enter a formal course in grades 9, 10, or 11, which involves proofs and problem solving; however, we should be realistic and state that not all learners with handicaps to learning and achievement will arrive at this point. It is not imperative that they do reach this point since we will have achieved the desired goals long before this and geometry is seen as a relief from working with numbers and not an end in itself.

DEVELOPMENTAL TEACHING OF MEASUREMENT

The Process of Measurement

The measurement strand of the mathematics curriculum is unique in that it is mainly concerned with a process—a unifying theme—that remains somewhat constant throughout the various content elements. This is a discrimination process in which the learner becomes involved in such activities as comparing, categorizing, and sequencing. In order to function or to be able to work within these activities the learner must possess what Piaget called the reversibility concept and be moving into the concrete operations stage of cognitive development. Learners with special needs, learners with handicaps or disabilities, may arrive at these stages later than other learners; therefore, we should work in the strands of geometry and whole numbers until a readiness has been developed for the process of measurement.

An understanding of the process of measurement can best be developed by going from nonstandard, arbitrary units to develop the process, to the need for a standard unit, and on to the development and use of standard units. This understanding is formalized in the formal stage of cognitive development; thus the process of measurement should be developed carefully over a long period of time. The prerequisites of specific vocabulary, general terminology, and experiences in comparing, categorizing, and sequencing should be developed carefully while learners are preoperational. The need for units and the use of arbitrary units occur as part of the natural development of the use of numbers. The final development involving linear measure follows a basic understanding of whole numbers, an introduction to fractions, and experiences with line segments. The most dominant method throughout this development should be a laboratory approach. Using different modes of instruction is as important in this strand as in other strands.

Measurement—A Connecting Link

The measuring process is a significant mathematical concept. Measurement is a function, a rule for operating. Everything from geometry concerns sets of points. Lines, line segments, and planes are sets of points. Everything involving numbers concerns sets of numbers. If we stop here we have two unrelated parts of mathematics. The measurement concept serves to connect these two parts of mathematics. To demonstrate this concept let's use a set of points, a line segment, and a set of numbers.

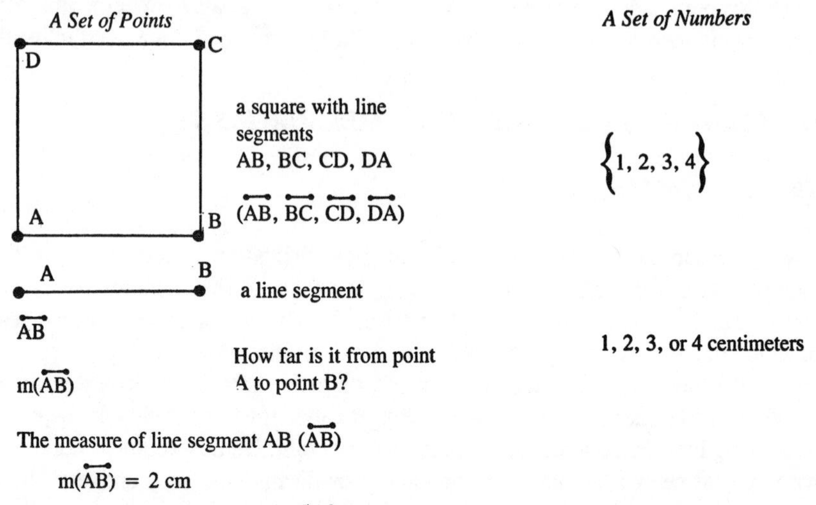

The measure of line segment AB(AB) "is equal to" 2 centimeters.

Measurement is a rule (a function) that is used to assign a number to the measure on a set of points. Assigning a number in this manner involves a measurement process. Often this rule means simply placing something in a proper category.

Linear Measure

With experiences with time and temperature and activities such as categorizing and sequencing, the learner should be ready to examine linear measurement. Such questions as How far? How long? How much? can now be answered. Linear measurement activities should be an extension of the relative size concepts developed in geometry and they should occur after the learner has had experience with whole numbers.

The developmental process should involve exploration, guessing or estimating, observing, and measuring for verification purposes. The following sequence of activities will serve to exemplify the process.

Introductory Activity

Have learners select some arbitrary length and use this length to determine the distance between two parallel walls of the classroom. Using only an element from the set of counting numbers, they should record their answers (1, 2, 3, 4, 5, . . .) in the following format:

The distance from wall "one" to wall "two" is ______________________________

____.

(Be sure to name your unit.)
Describe how you obtained your answer. ______________________________

____

____.

If your answer was more than one of your "units," how did you determine "how many?"

____.

If this is a class activity, discuss the results of various learners and decide upon a "best" answer for the distance between the walls.

Arbitrary Units

Any arbitrary unit may be selected to determine the distance between two points; however, each unit should be used in the same manner. One end of the arbitrary unit is placed at some starting point and the point where the other end falls becomes the starting point for the unit to be used again. The measuring process should continue along a straight line to determine the total distance between two points. These characteristics of the measuring process imply that measurement is a continuous operation, that is, one unit begins where the previous one ends and continues along the straight line.

Arbitrary units are used to give the learners the opportunities to develop their own unit for determining distances. Intuitively the need for a standard unit is to be established. Activities concerning arbitrary units can be introduced at all levels beginning in the enactive stage. At the preoperational stage the activities should be involved in building vocabulary with concrete objects, such as longer, shorter, taller; and in sorting objects with attributes that lead to length, area, and volume. At the concrete operations stage measurement becomes operational.

Similar procedures should be followed in introducing the standard units and in developing scales for measuring objects using these units. The concept of a unit or "what" the unit is should be developed before use is made of the unit.

AREA

The concept of area is mathematically similar to the concept of length in that some number is assigned to a set of points to name the amount of flat surface covered; however, this concept of area should come much later in the mathematical development.

Since area involves the ideas of multiplication and results in different kinds of units, the potential exists here for possible areas of difficulty for a slower learner or one with perceptual problems. Every opportunity should be taken to "show" the learner what square units mean. Cutting, drawing, and measuring activities are essential. Careful development from a square, to a rectangle, to other polygons should be followed using regions that have lengths and widths represented by whole numbers instead of fractions. The various formulas for area should be implied initially but not fully developed until the learner is beyond the concrete operations stage. Following the rectangle should be such figures as parallelograms so that learners can cut and construct rectangles (if necessary) to obtain the area.

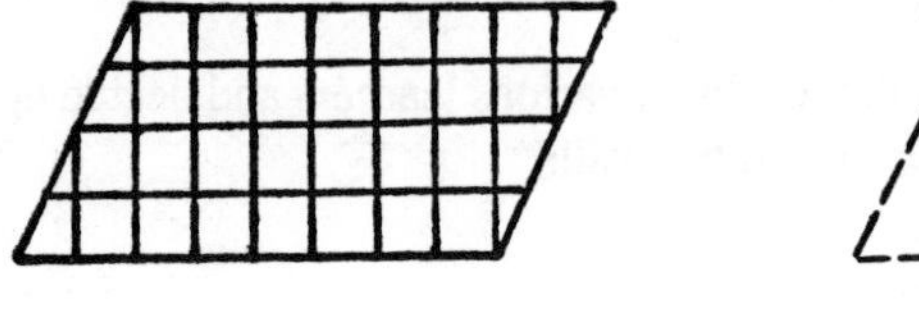

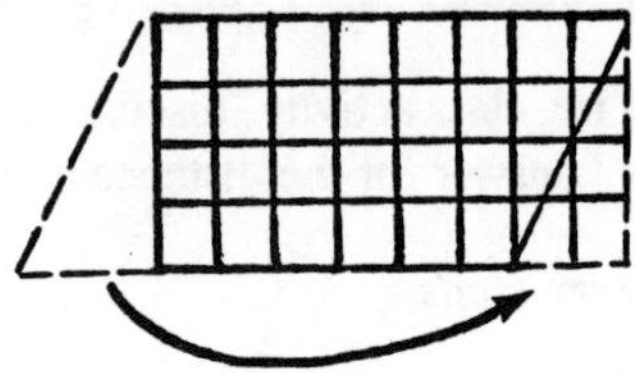

Area for triangles can be developed as an extension of this activity. The concept of congruence should be redefined: two figures (polygons) are said to be congruent if they fit exactly together when one is placed on top of the other or when all corresponding parts have the same measure.

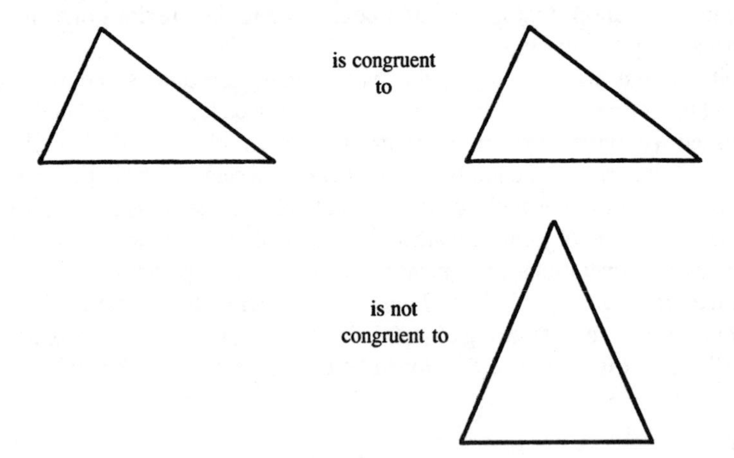

VOLUME

Volume is a more difficult concept than area but can be developed in a similar manner. The use of blocks to build volumes in a fashion similar to the use of area units as coverings for polygonal regions can strengthen the learner's intuitions for the unit and for congruence. Three-dimensional figures should be constructed to represent the figures already encountered in other geometric concepts. Learners should be expected to construct, identify, describe (state), and draw (graphically symbolize) three-dimensional figures. Using patterns is a good activity at this point.

MASS (WEIGHT)

Understanding mass is more complicated than understanding of length, area, and volume. Perception is the problem because holding two objects in your hands and saying which has the greater mass is difficult. Materials are necessary to enable the child to "see" whether two objects have similar masses (weigh the same) or not. A balance beam is essential. Objects placed on the ends of the balance beam weigh the same (have the same mass) if the beam is balanced (level with the floor).

The concept of equilibrium is introduced here and involves perception of the objects being weighed that is not directly related to the objects. As a result there may be some difficulty in examining mass (weight) in the following manner if we include it before the formal development stage. Learners should be given objects of pound and kilogram masses to handle and should be asked to weigh these objects and find other objects that are heavier and others that are lighter. Each learner should learn to read a scale to determine how much objects weigh. A good supplementary topic at this point is unit pricing, which involves weight.

If the metric system is being studied, the kilogram mass should be developed as follows:

1. Build a cubic decimeter from heavy cardboard (10 cm × 10 cm × 10 cm).
2. Secure the sides together.
3. Put a liner of plastic in the cube.
4. Fill the cube with water. This *defines* a liter of water.
5. Weigh the cube with the water. If you ignore the mass of the cardboard cube, the mass of the water is one kilogram.

If learners have been taught developmentally with teaching styles to match the learning styles, this development should continue through algebra, geometry, and advanced mathematics or through business mathematics, computer mathematics, or topics in mathematics. If learners have *not* been taught developmentally, there is a good possibility that there are mathematical gaps in their backgrounds. For learning-disabled individuals dysfunctions are magnified and we are faced with diagnostic teaching of mathematics before we can continue the usual content development.

DIAGNOSTIC TEACHING

Evidence to support the need for diagnostic teaching comes from studies of error analysis. The data (seen in Table 7–2) was collected from individuals enrolled in an alternative high school. Each error example was a systematic error involving a faulty algorithm. Faulty algorithms often go unnoted because the output behavior of the learner is in written form and among other data. An interview (a verbal question-and-answer period) is necessary to determine a learner's pattern of performance. Column 3 of Table 7–2 was completed in this manner.

While some in-class efforts have been attempted to meet the needs of various learners at the secondary level, few out-of-class efforts have been made. We have grouped for instruction, given individual attention in class (as the time allows), assigned workmates, and recommended tutorial help. When after-school work is undertaken to help learners with their assignments, we often repeat the classroom procedures. Some instructors do go to the point of giving differentiated assignments and quizzes to help learners to achieve; however, little in-depth diagnostic work has been attempted.

Underhill, Uprichard, and Heddens (1980) refer to a classroom diagnostic model. What we will be outlining is something beyond the classroom. To determine the strengths and weaknesses of learners in regard to mathematical concepts and skills, something more is necessary than our usual procedures. We need to know what algorithm(s) a learner uses to complete an example, exercise, or problem. We need to have someone trained to do a clinical mathematics interview. In an interview situation the strategies that a learner uses and any improper algorithms that have evolved can be discovered. It is here that we can begin to search for the proper modality in which to instruct a learner.

With experience a clinician will be able to prescribe the most direct route from problem error to defining difficulty to finding a successful mode of functioning. What appears to be important is to branch with the specific error as a base so as not to lose sight of the beginning.

At the secondary level, because of the maturity levels of the learners, teachers also need to be concerned with the kinds of objects and pictorial representations that are to be used. For example, a sixteen-year-old tenth grader in the alternative high school previously mentioned refused to work with any kind of objects because she felt they were "baby" materials (poker chips were being used). This girl was experiencing difficulty with regrouping (borrowing) in subtraction. She finally agreed to draw circles to represent the numerical values, but only if none of her classmates saw her doing the drawings, and soon learned to represent and regroup with her figures.

Essentially, the diagnostic procedure is a task-analysis procedure with the added dimensions of words, pictures, and objects. Many times a first-line effort is enough to get the learners back on track.

Table 7–2 Error Analysis, Grades 9–12

	Grade	*Age*	*Error*	*Explanation*	*Strengths*
1)	11	16	$\begin{array}{r}12\\+\ 5\\\hline 8\end{array}\quad\begin{array}{r}10\\+\ 7\\\hline 8\end{array}\quad\begin{array}{r}80\\+\ 4\\\hline 12\end{array}$	Learner added each single digit.	Learner knew basic facts.
2)	11	18	$\begin{array}{r}543\\-195\\\hline 452\end{array}\quad\begin{array}{r}621\\-335\\\hline 314\end{array}$	Learner always subtracted small number from large number.	Learner knew basic facts and idea of subtraction.
3)	10	17	$\begin{array}{r}754\\\times\ 79\\\hline 5286\end{array}$	$9 \times 4 = 36$, bring down 6 and carry 3. $7 \times 5 = 35$, $35 + 3 = 38$, bring down 8, carry 3, $7 \times 7 = 49$, $49 + 3 = 52$.	Learner knew basic multiplication facts and direction of multiplication algorithm.
4)	11	18	$\begin{array}{r}31\\7\overline{)217}\end{array}$ process error	The division was performed as $\begin{array}{r}1\\7\overline{)7}\end{array}$ and $\begin{array}{r}3\\7\overline{)21}\end{array}$ in this order.	Learner knew basic division facts.
5)	9	15	$\frac{2}{3} + \frac{3}{5} = \frac{5}{8}$	Learner added both numerator and denominator.	Learner added correctly.
6)	10	17	$\begin{array}{r}\frac{1}{4}\\\frac{3}{+4}\\\hline 12\end{array}$	Learner added all numbers.	Learner added correctly.
7)	10	16	$\begin{array}{r}\frac{9}{3}\\-\frac{7}{8}\\\hline \frac{2}{5}\end{array}$	Learner subtracted both numerators and both denominators (smaller from larger).	Idea of subtraction is present.
8)	10	15	$4 \times \frac{2}{3} =$ $\frac{4}{3} \times \frac{2}{3} = \frac{8}{9}$	Learner placed the denominator of the fraction under the whole number and then multiplied.	Idea of multiplication of fractions is present.
9)	12	17	$\frac{5}{6} \div \frac{1}{3} =$ $\frac{\cancel{6}^{2}}{5} \times \frac{1}{\cancel{3}} = \frac{2}{5}$	Learner inverted the ⅚ so that the ⅓ would divide.	Learner has the idea of division.

Example: 3.5 + 0.75 + 1.2 + .09
A learner will often add the numbers in decimal form and ignore the decimal in the answer or place it at random. A first-line effort might be to place the number values on a graph paper grid with the decimal point in one column.

3	.	5	
0	.	7	5
1	.	2	
	.	0	9

Zeroes may be added to complete the grid if this will aid the learner.

3	.	5	0
0	.	7	5
1	.	2	0
0	.	0	9

Some instructors use money values to assist the development.

The Goals of Diagnosis

The general goals of diagnostic teaching are to find a means by which a learner can function successfully and then to help the learner advance through these means to understand the standard algorithms and problem-solving procedures. Learning-disabled individuals must be helped to develop compensating skills. We must develop their strengths rather than dwell on their weaknesses. For instructors this may mean tape recording class presentations or notes in addition to providing handouts. Some learners may need to use tables for operations (e.g., the multiplication tables for whole numbers) or to use calculators as instructional aids. Others may need to have assistance in organizing materials. Once we find a functioning modality, we may need to enhance this modality instead of trying to change the learner to a more standard approach. If a learner needs and can use a calculator successfully, let's allow it to become not a crutch but a *tool*. Our greatest difficulty in diagnostic teaching at the secondary level appears to be changing the attitudes and instructional procedures of teachers.

ADVANCING DEVELOPMENTAL TEACHING

Individuals who are *ready* to begin the study of algebra will have developed an understanding of numbers in whole number, fraction, and decimal form through developmental and/or diagnostic teaching of mathematics. They will have explored geometry and measurement, and they will have experienced an introduction to problem solving. Thus, the secondary school mathematics program will be viewed as a continuing process from this initial development.

For learning-disabled individuals at the secondary level a mathematics program must be flexible. Algebra must *not* be viewed as a one- or two-year program, but as a continuous program to the level of understanding possible for a given group of learners. For learning-disabled individuals it is probably best to study algebra 1 and algebra 2 without a year break for geometry (as is the usual procedure). This program can be followed by the study of Euclidean geometry and analytic geometry. We will discuss both algebra and geometry in the following sections.

The Study of Algebra

For learning-disabled individuals the study of algebra must include a multimodal, stimulus-rich environment because of the demand to function in a symbolic mode. The program must have a definite structure, that is, with one section building upon another. (Some of the newer textbooks in this area lack such a precise structure.) The examples and exercises should be arranged to reinforce a concept or skill and extend to the next related concept or skill. Moreover, proper sequencing of ideas can lead to a discovery method of teaching which, in turn, should allow a learner to *experience* mathematics. The outline presented below will be followed by some examples of developmental strategies.

An Outline for Algebra

1. Numbers of Arithmetic
2. Real Numbers
3. Operations and Inverses
4. The Language of Algebra
5. Equations and Inequalities
6. Introduction to Proof
7. Simplifying Algebraic Expressions
8. Solving Equations, Inequalities, and Problems
9. Fractions
10. Fractional Equations
11. Algebraic Manipulation

12. Graphing
13. Relations and Functions
14. Linear Functions
15. Introduction to Quadratic Functions
16. Systems of Equations
17. Trigonometric Functions, Solutions of Triangles
18. Exponentials and Radicals
19. Quadratic Functions
20. Rational Functions
21. Complex Numbers

Factoring

Some authors list separate sections on factoring, but this outline contains factoring within such topics as solving equations.

The process of factoring provides an excellent opportunity to demonstrate developmental teaching. Provide the learners with pieces of oaktag in the format described in Figure 7–6. Actually, wooden models could also be used. Include several different sizes to enhance the concept being developed.

Ask the learners to determine if they can arrange the pieces of oaktag to form a square region. The learners should be able to construct such an arrangement. One possible arrangement is seen in Figure 7–7. The dimensions are $x + 1 + 1$ or $x + 2$ by $x + 2$, which produces $x^2 + 4x + 4$ if the lengths of the two sides are multiplied. Each element of this expression is a portion or portion of the pieces; therefore, we have a perfect square trinomial, $x^2 + 4x + 4$, with factors, $(x + 2)$ and $(x + 2)$. The learner now has a factoring model for this set of algebraic expressions in addition to the verbal description. Every opportunity should be used to build models of the algebraic situations since they serve to clarify and help answer the *why* question.

Often, our developmental strategies exist within the numbers and their patterns. Suppose, for example, that we want to convince a learner that a negative real number "times" a negative real number equals a positive real number. The following pattern might be employed.

1. Let $+3 \times 0 = 0$.
2. Let $0 = (+2 + -2)$.
3. Equation one becomes $+3(+2 + -2) = 0$.
4. $+3(+2 + -2)$ becomes $(+3 \times +2) + (+3 \times -2)$.
5. Equation three becomes $(+3 \times +2) + (+3 \times -2) = 0$.
6. We know that $(+3 \times +2) = +6$; therefore, $(+3 \times -2)$ must be equal to -6 to make our sentence true.
7. Now, let the example be $-3 \times 0 = 0$.

Figure 7–6 Algebraic Pattern Pieces

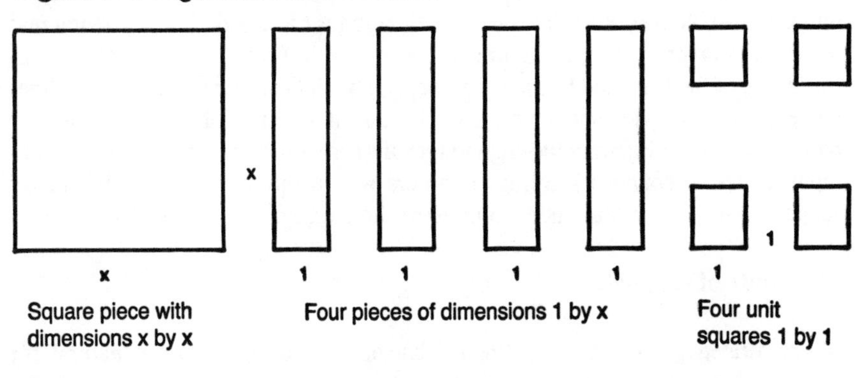

Figure 7–7 Geometric Representation of a Perfect Square Trinomial

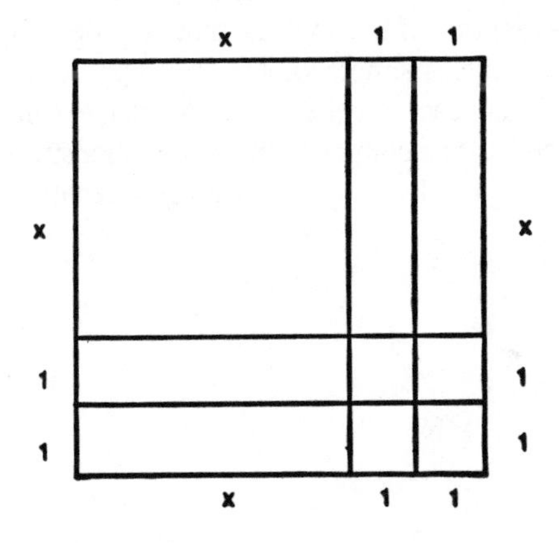

8. Again, let $0 = (+2 + -2)$.
9. Equation seven becomes $-3(+2 + -2) = 0$ or $(-3 \times +2) + (-3 \times -2) = 0$.
10. We know from sentence six that a negative number "times" a positive number equals a negative number; therefore $(-3 \times +2) = -6$.
11. This forces (-3×-2) to equal $+6$ if the equation in statement nine is to be true.

This pattern involves a distributive property for real numbers and presents a logical pattern. Several such patterns exist within the structure of algebra and they should be used to enhance understanding of concepts.

Graphing solution sets also provides another way to enhance learning and to introduce problem solving. The general equation for a straight line involving real numbers as ordered pairs of the line is $y = mx + b$ where x and y vary while m (the slope of the line) and b (the y-intercept) remain the same within a given line. Understanding is evident when the learners can tell the effect of changing m and/or b in an equation. Graphing the changed results serves to provide another model or input device to enhance learning. Learners with disabilities need such a broad scope to gain a complete understanding of the concept.

The Study of Geometry

The strategy in teaching geometry that appears to be the most natural for learning-disabled individuals is transformational geometry taught from a laboratory approach. In such an approach each concept (with the exception of deductive and inductive proofs) is developed from models using wax paper, graph paper, and paper to be folded by precise rules. For example, angle bisectors can be constructed by folding a diagram of the angle so that the sides of the angle touch. The crease is the angle bisector. A simple knot flattened produces a regular pentagon. Congruent figures overlay each other exactly. A triangle can be drawn and angle sections cut and placed with a common vertex to demonstrate that the sum of the interior angles of a triangle is 180 degrees. A suggested outline for geometry using this strategy follows.

An Outline for Geometry

1. Informal Geometry
2. Segments
3. Angles
4. Triangles
5. Parallel Lines
6. Polygons
7. Similar Polygons
8. Circles
9. Area

Under a developmental approach each idea is built, described, classified, and used to enhance understanding. For example, three wooden rods or straws of equal length are joined together to form a triangle (an equilateral or equiangular triangle). A pattern is traced on paper for the triangle. From this pattern other copies are made for folding. Discoveries such as the following can be made: (1) all angle bisectors pass through the same point, (2) all medians pass through the same point, (3) all altitudes pass through the same point. One side (a line segment) of the triangle can be translated (moved) to form a square or rectangle (see Figure 7–8). From this a pattern can be created to investigate the properties of the figure. Translating the other sides of the triangle can produce a pattern for a triangular prism (see Figure 7–9).

Figure 7–8 Translation of a Line Segment

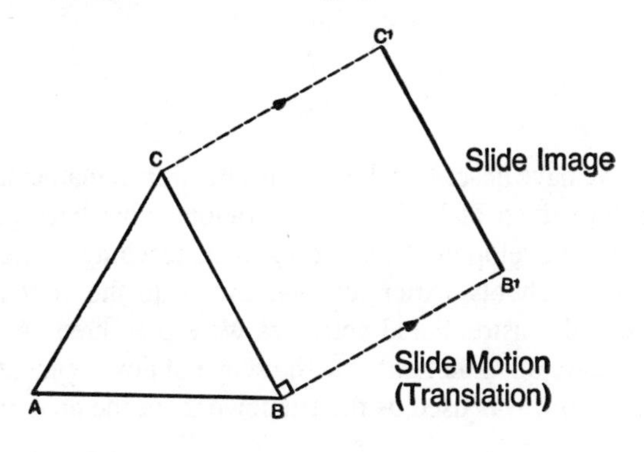

Figure 7–9 Pattern for a Triangular Prism

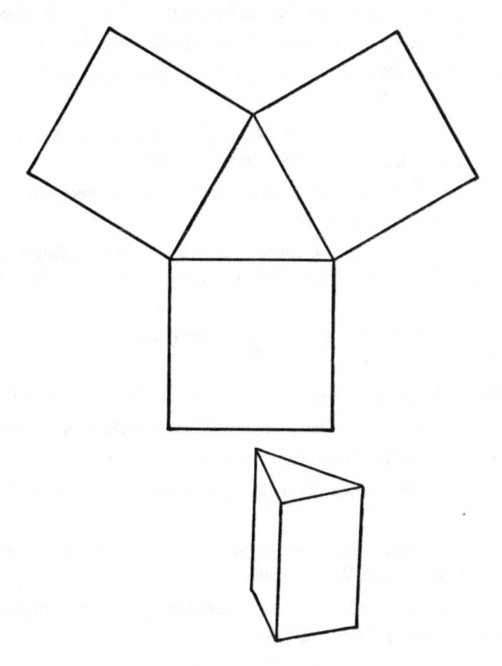

Problem solving as a proof can be introduced when we ask "Will the properties be the same if we change the triangle?" This expanding, related development can be most rewarding because both teacher and learner are creating mathematics. Some basic ideas for such a development may be found in the *Motion Geometry* set of books by Phillips, McKeeby, and Zwoyer (1969) and the extension to tenth grade geometry may be found in Coxford and Usiskin's *Geometry: A Transformational Approach* (1971).

SUMMARY

In this chapter we have described the current situation in mathematics education and the relationships to special education. A rationale has been presented along with an overview of developmental and diagnostic teaching in the middle/junior high and secondary schools. Attention was drawn to the content in grades 4 through 8 and sample instructional activities were presented. A wide range of strategies were illustrated to demonstrate the scope of developmental teaching. A task-analysis procedure was used as the framework for the chapter.

REFERENCES

Bloom, B.S. (Ed.). (1968). *Taxonomy of educational objectives*. New York: David McKay Co.

Braunfeed, P.G. (1969). *Stretchers and shrinkers* (Books 1–4). New York: Harper & Row.

Buswell, G.T., & Clark, W.W. (1925). *Buswell-John diagnostic test for the fundamental processes in arithmetic*. Indianapolis: Bobbs-Merrill.

Cawley, J.F., Goodstein, H.A., Fitzmaurice, A.M., Lepore, A., Sedlak, R., & Althaus, V. (1976). *Project MATH*. Tulsa, Okla.: Educational Development Corp.

Commission on Mathematics of CEEB. (1959). *Program for college preparatory mathematics*. New York: College Entrance Examination Board.

Conference Board of the Mathematical Sciences National Advisory Committee on Mathematics Education (NACOME). (1975). *Overview and analysis of school mathematics, grades K–12*. Washington, D.C.: NACOME.

Coxford, A.F., & Usiskin, Z.P. (1971). *Geometry: A transformation approach*. River Forest, Ill.: Laidlaw.

CTB/McGraw-Hill. (1975). *Diagnostic mathematics inventory*. Monterey, Calif.: Author.

Ingram, C. (1968). *Education of the slow-learning child*. New York: Ronald Press.

Inskeep, A. (1938). *Teaching dull and retarded children*. New York: Macmillan.

Kline, W.E., & Baker, H.J. (1963). *Bobbs-Merrill arithmetic achievement tests*. Indianapolis: Bobbs-Merrill.

Meyer, E. (1972). *Developing units of instruction: For the mentally retarded and other children with learning problems*. Dubuque, Iowa: W.C. Brown.

National Council of Supervisors of Mathematics. (1978, February). Position statement on basic skills. *The Mathematics Teacher, 71*(2), 147–152.

National Council of Teachers of Mathematics. (1970). *Experiences in mathematical ideas* (Vols. I & II). Washington, D.C.: Author.

National Council of Teachers of Mathematics. (1978, February). Position statement on basic skills. *The Mathematics Teacher, 71*(2), 147.

Phillips, J.M., & Zwoyer, R.E. (1969). *Motion geometry* (Books 1–4). New York: Harper & Row.

Secondary School Curriculum Committee of NCTM. (1959). *The secondary mathematics curriculum.* Washington, D.C.: National Council of Teachers of Mathematics.

Taylor, R. (1978, February). The question of minimum competency as viewed from the schools. *The Mathematics Teacher, 71*(2), 88–93.

Underhill, R.G., Uprichard, A.E., & Heddens, J.W. (1980). *Diagnosing mathematics difficulties.* Columbus, Ohio: Charles E. Merrill.

Chapter 8

Mathematics in the Real World

Mahesh C. Sharma

More than three centuries ago Francis Bacon said, ''Many parts of nature can neither be invented with sufficient subtlety, nor demonstrated with sufficient perspicuity, nor accommodated with sufficient dexterity without the aid and intervening of mathematics; of which sort are perspective, music, astronomy, cosmography, architecture, engineery and divers others. I may make only this prediction, that there canst fail to be more kinds of (mathematics) as nature grows further disclosed.''

Today's world, which Bacon so sagely intuited, does indeed include mathematics of various kinds and levels: the mathematics of invention in the fields of science and technology; the mathematics of understanding and exploration of nature; the mathematics of day-to-day living, and finally the mathematics of recreation and pleasure. While it is clear to everyone that mathematics is essential in such activities as engineering, building, and developing and testing theories regarding physical phenomena, we are not always so cognizant of our dependence on the ability to do mathematics in numerous aspects of our daily lives. This is evident from the following example.

Erik F. came to the office of a professor of mathematics with a request for help in developing his mathematical abilities. When asked what type of mathematics he needed and why he thought he needed it, he gave this explanation. He said he had always had difficulty with mathematics from elementary school through high school. When he had graduated from high school he had been glad that he did not have to take any more mathematics courses and pleased that he had been successful in ''bluffing'' his way through high school mathematics. After high school he had enrolled in the university and carefully avoided all contact with mathematics. Again he had been happy that mathematics ''was no longer in his life.'' A year later he had dropped out of the university. Now for the last few years he has been an interior decorator and also restores old houses. He enjoys his work. He is an avid sportsman and particularly enjoys playing darts with his friends. He has several friends from the high technology field.

Recently when he was playing darts he realized that his friends could count, add, and subtract in their heads; he had difficulty keeping up with them and was embarrassed by not being able to do so. That is when he began to see that mathematics was not going to "leave him alone" even in the "real world." He also started thinking about the impact that mathematics was having on his business and his relationship with his friends. He found that he postpones many tasks, particularly those that involve calculations and estimates, and many times procrastinates in submitting bills and estimates for jobs, all because he wants to avoid facing the mathematics involved in the process. He even avoids participating in discussions that his friends have because he feels he does not possess the reasoning and logic needed for those discussions. It is his belief that his friends have a higher order of reasoning and logic because of their training in mathematics. He is beginning to feel that his profession, his recreation and pleasure, and the world around him all deal with mathematics to a certain degree. He strongly feels that he needs a certain degree of proficiency in basic mathematics.

Erik has thought of the alternative: using the calculator, for figuring out estimates for jobs, materials to be used, or time spent on projects, for percentages of profit, commissions, discounts, and variations. He has found that even to use the calculator effectively he needs specific methods and understandings. Many times the answer that the calculator gives does not look "right," but he does not know how to check it or rectify the error.

The problem has now reached the stage that it is beginning to undermine his capability as a successful interior decorator and businessman. Erik's self-concept is beginning to be affected by his inability to do the mathematics involved in his work. He has also explored the alternative of hiring an assistant to do the calculations, but he knows that in every business decision he needs some mathematics and now even to function effectively in his personal life he can no longer manage without it.

Erik is not alone. Most people need to have proficiency, at least in the mathematics they learned in high school, in order to solve "real world" problems.

The language of number, of quantity, measurement, and comparison has become so much a part of everyday life that we are hardly aware of it. When we speak of "fifty-fifty," for example, we do not realize that we assume a knowledge of percentage. A glance at a daily newspaper will verify how important to every citizen, irrespective of the requirements of his or her occupation, a knowledge of mathematics may be to understand what is going on in the world.

At different levels mathematics aids us in various ways. The basic understanding of the world we live in involves concepts of quantity, space, time, and their interrelationships, which are so fundamental that one finds their presence in all situations. For example, there are many phenomena in the physical world, society, and in human nature that are affected by time variables and chance factors. It is important for every adult to know that there are laws of chance, and to understand what conclusions one can draw when one is faced with uncertainty (probability and possibility). The step from description of observations of a phenomenon and analysis of data to mathematical models for prediction in the decision-making process is a significant one. Understanding chance phenomena is fundamental in the study of natural and social sciences but in addition there is hardly any game of recreation where the outcome is not affected by probability and chance. Some familiarity with statistical thinking is essential for everyone.

Making decisions intelligently—how to pick the best among the alternatives available—this too is a requirement of daily living. To give a simple example, suppose I wish to fence a rectangular plot so that it will have an area of 100 square yards. Fencing costs $8.00 per yard. What shape should I make the plot?

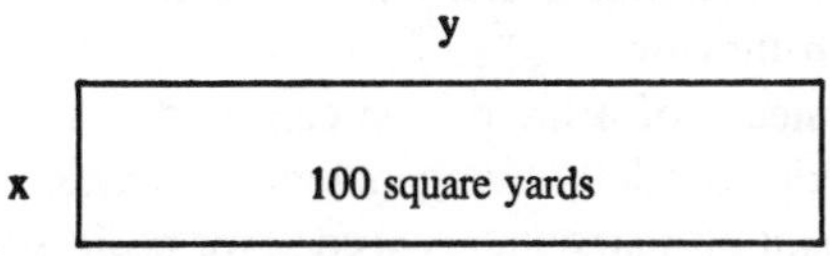

y

x | 100 square yards

If I make it 20 yards wide, it must be 100/20 or 5 yards long so that the area will be 20×5 or 100 square yards. Then I need $20 + 5 + 20 + 5$, or 50 yards of fence, which costs $400. If $x = 10$, then $y = 100/10 = 10$, and the fence costs $320. If $x = 25$, then $y = 100/25 = 4$, and the fence costs $464. The answer can be obtained by applying some mathematics to the equation: cost $= 8 \times 2 (x + 100/x)$, which shows that $x = 10$ is optimal.

In order to predict the consequences of the choices before us, we must have a model of how the world works. If we know the outcome of all alternatives and how to evaluate them, we can apply mathematics to decide which choice is best. The situations where an optimal choice is to be made are in abundance in our daily life.

It is important to know some applications and applicability of mathematics but it is equally important to know when a particular application is possible. We must understand both the power and the limitations of mathematics—how to use it to make the best choice when it is applicable and what assumptions underlie any given application of mathematics. Most people need to know some facts about numbers and space in order to function in the modern world. The equations in a pamphlet for farmers and the graphs in the daily newspapers indicate something about the degree of mathematical literacy necessary today for almost everyone.

PUTTING MATHEMATICS TO USE

There are two kinds or levels of mathematics that are needed in the real world: (1) higher mathematics and (2) arithmetic, simple algebra, and geometry. Higher mathematics involves calculus, finite mathematics, probability and statistics, applied algebra, analysis, differential equations. This type of mathematics is involved in a diversity of professional applications: engineering, medicine, the physical and natural sciences, social science research, archeology, anthropology, and even languages. For example, the "Homeric question"—namely, whether *The Iliad* represents the work of a single man (Homer) or a collection of orally transmitted hero legends—has received its definite solution by a young American philologist. With the help of mathematical statistics and a computer, he determined that it is not an anthology but was created by a single poet.

As for arithmetic, algebra, and geometry, they are all about us. Geometry, which deals with form and shape or "spacial relationships," continually employs arithmetic and algebra. Trigonometry is a special kind of mathematics that employs all three. For almost every practical problem outside the bounds of simple trading calculations mathematical analysis and solutions employ algebra and geometry as well as arithmetic.

The mathematical needs of a lay person can be divided into two categories: mathematical knowledge for daily living and mathematical insight for recreation and leisure. The amount of mathematics needed for professional purposes varies so enormously that hundreds of thousands of books are written on the subject of higher mathematics. Their contents vary from simple arithmetic for a real estate agent to partial differential equations for electronic engineers. Here we want to discuss the mathematics needed for daily living in the "real world" by almost everyone. In the following sections, we will identify some of the common but important concepts of everyday mathematics.

Living within One's Means

One of the first mathematical problems that affects every individual is that of accounting for one's own income—knowing the constraints and limitations of your means. If you spend without planning and budgeting, you should be prepared for frequent shocks. Very few people have detailed accounting methods and ledgers of their income and its disposal but most have some idea of estimating expenditures in advance, and many people, especially in these days of rapid variations and escalations in the cost of living, keep household accounts.

This activity calls for the skills of:

1. estimating
2. basic operations of addition, subtraction, multiplication, and division of whole and decimal numbers
3. percentages and simple fractions
4. extrapolation

Estimating

In real life, in day-to-day living, we use estimation more often than actual calculations. An important and sometimes neglected component of school teaching is the estimation aspect of quantitative work. When we say the answer to 12×18 is more than 100 and close to 200, we are estimating. Estimating the size of an answer is important because this sort of approximation may be good enough for the problem we are solving or the estimation might serve as a good check on the exact answer whether produced by man or machine. Machines seldom make errors in simple calculations of the type $3 + 3 = 6$; their errors tend to produce answers absurdly wrong, which an estimation would check. Estimation procedures become even more important now that pocket calculators and minicomputers are beginning to be commonplace items.

The most common and popular meaning of estimation is in finding approximate answers to numerical problems, where the approximation is used to generate a "rough" answer by translating the numbers in the original problem into "rounded off" figures. For example: John wants to carpet his family room. The room is 11½ feet long by 9¾ feet wide. How much carpet should he buy? Since in carpeting the room there will be cutting and some wastage involved, he approximates the numbers (12 feet long and 10 feet wide) by "rounding off" and finds the answer by multiplying the new numbers.

In estimating "rounding off" of numbers is a key skill. If the number in the digit immediately to the right of the place being rounded off is more than 5, then the number in the intended place is increased by one more; otherwise it remains the same.

In actual problems, we can arrive at estimated solution through two ways:

1. The *benchmark technique*. Based on previous experience with a similar problem the person makes an estimate; in essence it is dependent on comparisons made with a known standard and validated experience. For example, How high is the top of the bulletin board in church? To answer this question a person might use the following process. The top hits my upstretched arm, and I know I can reach up about 7½ feet, so the top of the bulletin board is about 7½ feet high.
2. *Decomposition/recomposition*. For many estimation problems, no appropriate benchmark may be available. Then the estimation process involves a decomposition of the problem into smaller problems that can be estimated by the benchmark techniques. After finding the benchmark estimation for the individual smaller problems one recomposes the estimates into an estimate of the original problem. For example, a man may need to find the height of a building but he may not have had any experience that could give him benchmarks or standards for comparison. He can decompose the problem

> into estimating the height of each story. Using his own height he can estimate that the ceiling of each story is about eight feet. Allowing two feet between stories for pipes, wires, insulation, ducts, etc., the total estimate of a single story comes to about eleven feet. Suppose there are fifty stories in the building, this means that the total height is 50 × 11 or about 550 feet altogether.

Learning to estimate—learning the scales and dimensions by which we make quantitative judgments about the environment and appropriate action in that environment—is an important mathematical skill. Estimation is indispensable even to measurement, since one must estimate to choose appropriate units of measurement and to determine whether results are reasonable.

Interest

We have some money that we do not wish to spend at the moment. We put it in the bank or invest it for a return of income on it. Interest is usually expressed as a rate percent per annum. If the rate is, for example, 6% per annum, this means that if you deposited the money or lent someone $100.00 for a year, you would receive $106.00 back, the additional $6.00 being the interest. If you had left the money on loan for two years, on this basis, you would be entitled to $12.00 interest.

This assumes that the interest is not asked for until the end of two years. If, in the case of the $100.00 loan for two years at 6% per annum, the interest had been paid by the year, there would be $106.00 owing at any time during the second year. As 6% of $106.00 is $6.36, at the end of two years you would be entitled to $112.36, not $112.00. This is called compound interest, the additional $.36 is interest on the first year's interest, which is put in and regarded as part of the loan.

All adults deal with money and interest and therefore must understand the difference between various rates and types of interest to make the best use of their resources. For loans, credit buying, time payments, and mortgages, we can't do without a fundamental understanding of the following skills:

1. percentages
2. estimation
3. basic operations of addition, subtraction, multiplication, and division of whole and decimal numbers
4. comparison
5. extrapolation

Percentages

Percentage is a particular form of fraction in which the denominator is always one hundred. Percent means per hundred (cent). Thus 5% means 5 percent or 5 per hundred or $\frac{5}{100}$. The important thing to understand is that 100% (one hundred percent = 100 per 100 = 100/100 = 1) is 1 and that you may therefore multiply by 100% without changing the value. In order to express any fraction or decimal as a percentage, we multiply it by 100%, as in the following example:

$$\frac{1}{4} = \frac{1}{4} \times 100\% = \frac{100}{4}\% = 25\%$$

$$\frac{1}{2} = \frac{1}{2} \times 100\% = \frac{100}{2}\% = 50\%$$

$$\frac{1}{5} = \frac{1}{5} \times 100\% = \frac{100}{.5}\% = 20\%$$

$$\frac{9}{10} = \frac{9}{10} \times 100\% = \frac{900}{10}\% = 90\%$$

$$\frac{1}{3} = \frac{1}{3} \times 100\% = \frac{100}{3}\% = 33\tfrac{1}{3}\% = 33.3\%$$

$$.25 = .25 \times 100\% = 25\%$$

When we multiply a decimal by 10 the decimal moves to the right by one place, when we multiply by 100, the decimal moves to the right by two places. Table 8–1 shows some fraction, decimal, and percent equivalents.

The following rules are guidelines for making conversions:

1. From fraction to decimal divide the numerator by the denominator.
2. From decimal to percent move the decimal point two places to the right.
3. From percent to decimal move the decimal point two places to the left.
4. From fraction to percent multiply the fraction by 100 percent and simplify.
5. From decimal to fraction express the decimal as a fraction and reduce it.

Profit and Loss

The definition of profit and loss is simple: if the selling price minus the cost price is more than zero there is a profit, if it is less than zero there is a loss. The problem in calculating net profit involves a good deal more than subtracting cost price from selling price if the true cost is to be considered. Many other factors are to be taken into account. There are all the selling expenses, advertising, and the cost of

Table 8–1 Fraction, Decimal, Percent Equivalents

Fraction	$\frac{1}{10}$	$\frac{1}{5}$	$\frac{1}{4}$	$\frac{1}{8}$	$\frac{3}{4}$	$\frac{9}{10}$	$\frac{1}{3}$	$\frac{2}{3}$
Decimal	0.1	0.2	.25	0.125	0.75	0.9	0.333	0.666
Percent	10	20	25	12.5	75	90	33.3	66.6

upkeep, rent, wages, lighting and heating, etc. According to the Small Business Administration, more small manufacturers and businesses and self-employed people fail because they fail to keep good records and apply these mathematical considerations.

Hire and Purchase

Most of us at some point in our lives rent cars, furniture, equipment or lease machines either for personal, professional, or business uses. A comparison of different prices and rates involves mathematics. For example: the AA Car Rental Company charges a daily rate of \$25 and 25¢ per mile after the first 100 miles. BB Car Rental Company charges a daily rate of \$20 and 20¢ per mile. If you used the car for 600 miles in two days which one is the better buy? If you traveled 100 miles in a day, which one would be the better buy then? The mathematics involved in this type of problem is:

$$\text{Charges} = \text{daily rate} \times \text{no. of days} + (\text{no. of miles} - \text{free miles}) \times \text{mileage rate}$$

In calculating AA Car Rental Company charges we have:

$$\text{Charges} = \$25 \times 2 + (600-100) \times 25¢ = \$50 + \$125 = \$175$$

In calculating BB Car Rental Company charges we have:

$$\text{Charges} = \$20 \times 2 + (600 - 0) \times 20¢ = \$40 + \$120 = \$160$$

The arithmetic is simple; the main thing is setting up the equation.

Taxes

Almost every adult in the United States is a taxpayer. The taxpayer will wish to know how his or her taxes are being calculated, if only to make sure the amount is not too much. You can check the tables that are worked out for different income levels that are available in every income tax office and public library. An understanding of the following procedures will make preparing your tax returns a much easier, routine task.

Taxpaying is an individual concern, but we are citizens in a country where our taxes are contributed in return for services and protections. We all have a stake in the political and economic decision-making process that takes place at local, state, and national levels of government.

When we read that the total federal budget is $648,000,000,000 and that of this 12 percent, or $77,000,000,000, is collected in the form of direct taxes on personal income and a deficit (expenditure is more than the income) of 140 billion dollars remains, a great deal of understanding is required.

How big is a billion? How big is a million? What does this deficit mean? When the interest rates go up and down what effect does it have on the consumer? What is the effect of high interest rates on individual mortgages? What is the effect of deficits on individual incomes and living standards? We apply our understanding of such matters when we go to the polls, choosing people to represent our choices in the day-to-day management of affairs that affect us all. Meaningful participation cannot exist without a foundation in mathematics.

Travel and Transportation

Americans travel more than any other group of people in the world. That involves some doing—both budgeting and calculations. The mathematical concepts involved in the pursuit of travel and transportation include the idea of distance, direction, map and graph reading, map making, measurement of time and speed, costs, and the interrelationships of these factors with each other. The complexity of the mathematics involved depends upon whether the person is the user of the information or generator of the information. For a professional who generates the information the mathematics needed falls in the category of higher mathematics, whereas for the user the mathematics involved could be only simple arithmetic and simple algebra. The equation that relates time, distance, and rate is $d = rt$. Cost = rate per mile × distance.

Games and Chance

Almost every game, whatever the degree of skill it demands of the players, has its own elements of chance. The part that chance plays in the final result varies enormously. It is these elements of chance, this infinite variety of circumstances that may arise by accident, that forms part of the appeal of many games. Were it not so, it would be useless for players of unequal skills to compete with each other, since the result would be a foregone conclusion.

While it is possible in some games to calculate the chances of certain situations arising, or of a particular player being successful, in others no amount of calculation can estimate the effects of chance. Calculating odds for the winning of a particular team, getting a particular card or groups of cards in a bridge or a poker game, winning a lottery, getting a particular number in the game of roulette, etc., involves some understanding of permutations and combinations. But the calculation of chance factors and their occurrences is not limited to games only. When we talk of a 50 percent chance of rain and 20 percent probability of one's selection to a task, we are talking of calculating chance or the probability of that happening.

Tables and Graphs

There are many instances where some sort of calculation is repeated often and it is a great saving of time to have a set of answers worked out, carefully checked and neatly tabulated for future use. Banks, supermarket checkout counters, sales clerks, physicians, and many others have ready-made charts, graphs, and tables for ready reference. Newspapers, propaganda materials, and annual reports from corporations and organizations, and many other individuals and groups present information through graphs (bars, histograms, pie charts, tables) and other visual materials. The utilization of these visual aids to information requires mathematical expertise ranging from understanding simple numerical tabulations to percentages and ratio and proportions.

Mathematics for the Handyman

A large majority of Americans like to do things around the house that involve some carpentry, landscaping, plumbing, and electrical work. There is a special kind of calculation widely used in most stages of construction and maintenance work that requires a knowledge of fractions. In fact, all household activities at one time or another will involve the following mathematical skills:

- basic operations of addition, subtraction, multiplication, and division of whole numbers, fractions, decimals, and signed numbers
- estimating
- working with geometrical shapes and interrelationships
- calculating areas, volumes, weights, and surfaces
- calculating ratios, percentages, averages, and proportions
- solving mixture problems
- isolating variables and understanding the mutual effects of variables and their individual contributions to problems or solutions
- reading meters, bills, and reports relating to specifications

More specifically, one needs an understanding of:

- pressure per load for foundation work
- volume, area, and weight for layout work, painting, etc.
- ratios (cement, water, sand, gravel, paint, or paint thinner)
- spacing for brickwork, tiling, flooring, etc.
- stress for beams, posts, spans, and arches

Geometry: Lines, Shapes, and Angles

Geometry has been defined as the investigation of space. It involves the study not only of solid objects but also of flat surfaces of all kinds. A knowledge of geometry is a practical necessity. With some elementary drawing instruments—ruler, compass, set squares, and protractor—many things can be discovered and many obvious facts verified. Looking at the hands of a clock or watch, the spokes of a wheel, criss-crossing of lines, or the legs of a pair of compasses one can see angles. An angle is formed when a straight line turns about a point in its own length; the amount of turn will determine the amount of angle. For example, in one hour the minute hand of a clock makes one complete revolution and the angle turned through is 360 degrees. A minute hand in 15 minutes will make an angle of 90° with the original position. When the sum of two angles that stand side by side is 180° they are referred to as supplementary angles; each is called the supplement of the other. Two angles that together make one right angle are called complementary; each is the complement of the other. Each pair of angles formed when two straight lines cross are called vertically opposite angles; vertically opposite angles are always equal. When two lines are perpendicular to one another they form a right angle (90°). Almost every craftsman deals with right angles. It is often important to ensure that a surface is horizontal or vertical or that a corner is square. For that purpose, a builder uses a spirit level, a plumb line, or a set square. A draftsman uses a T-square and set square.

The triangle has three sides; these may all be of different lengths. There are different kinds of triangles; the scalene has unequal sides and angles, the right-angled triangle has one angle of 90 degrees. The relationship between the sides in a right-angled triangle is defined in the Pythagorean Theorem as: $(\text{hypotenuse})^2 = (\text{base})^2 + (\text{altitude})^2$. In other words, if we know two sides of a right triangle, we can calculate the third one. For example: in triangle ABC, B is 90°, AC is the hypotenuse, BC is the base, and AB is the altitude. Using $(AC)^2 = (BC)^2 + (AB)^2$, let us assume that $AB = 3$ and $BC = 4$, then $(AC)^2 = 3^2 + 4^2 = 9 + 16 = 25$. $AC = 5$ units.

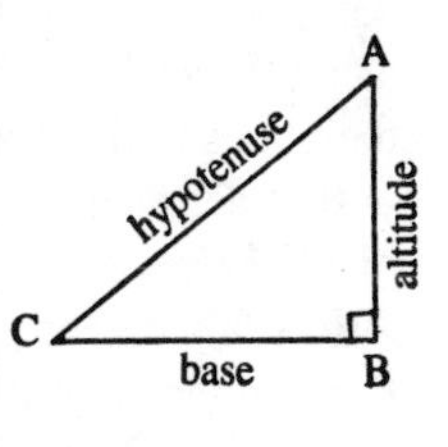

Another use of triangles is in scale drawings, enlargements, contractions, and map making. When two triangles have the same shape, corresponding angles are equal, and if one triangle is a larger version of the other, such triangles are said to be similar. The principle of similarity is not restricted to triangles. If, for example, we examine a specimen under the microscope the image seen is similar in every way to the original object. Everytime an architect or engineer makes a scale drawing, or a photographer makes an enlargement, or we use field glasses or a telescope, this principle is realized. For example, the triangles ABC and XYZ are similar.

AB↔XY, BC↔XZ, AC↔YZ

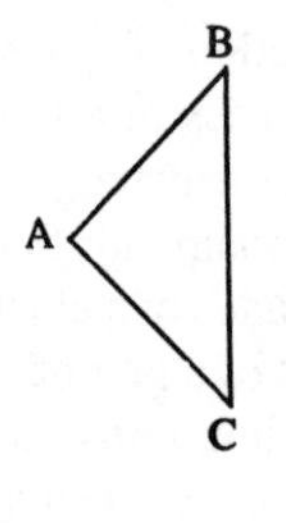

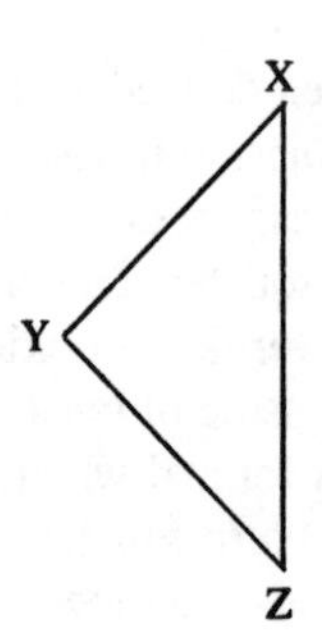

Then $\frac{AB}{XY} = \frac{BC}{XZ} = \frac{AC}{YZ}$

If we know the dimensions of one triangle and the scale of transformation, the dimensions of the other triangles can be calculated.

Signed Numbers

The plus and minus signs in arithmetic indicate operations of addition and subtraction respectively, but in algebra $-$ is regarded as the reverse of $+$. Any number with a $+$ before it is a gain, increase, asset, or lies on the right of zero on the number line, while any number with a $-$ sign before it is a loss, decrease, debt, or lies on the left of zero on the number line. If $+$ indicates a measurement along the scale in one direction, then $-$ indicates measurement along the scale in the opposite direction.

If $+17°$ means a temperature of 17° above zero, the $-17°$ means a temperature of 17° below zero.

If $+500$ means a deposit to an account (credit), then -500 means a withdrawal from the account (debit).

If 1,500 means 1,500 feet above sea level, then $-1{,}500$ means 1,500 feet below sea level.

If $+568$ means A.D. 568, then -568 means 568 B.C.

If $+10$ means 10 paces forward, then -10 means 10 paces backward.

These numbers with a sign attached are called directed or signed numbers. Consider the following examples:

(a) $+6 + 2 = ?$	(e) $+6 \times +2 = ?$	(i) $+6 \div +2 = ?$
(b) $+6 - 2 = ?$	(f) $+6 \times -2 = ?$	(j) $+6 \div -2 = ?$
(c) $-6 + 2 = ?$	(g) $-6 \times +2 = ?$	(k) $-6 \div +2 = ?$
(d) $-6 - 2 = ?$	(h) $-6 \times -2 = ?$	(l) $-6 \div -2 = ?$

In examples (a) and (d) both numbers have the signs indicating the same activity; in such a situation we apply the following rule:

Rule 1. For the same signs we add the numbers and use the common sign.

Thus $+6 + 2 = +8$; $-6 - 2 = -8$.

In examples (b) and (c) the two numbers have opposite signs indicating opposite activity. In such situations we apply the following rule:

Rule 2. In the case of opposite signs we subtract the numbers and use the sign of the larger number.

Thus $+6 - 2 = +4$; $-6 + 2 = -4$.

In multiplication and division we have the following rules:

Rule 3.

$$\begin{array}{l} + \times + = + \\ + \times - = - \\ - \times - = + \\ - \times + = - \end{array}$$

Rule 4.

$$\begin{array}{l} + \div + = + \\ + \div - = - \\ - \div - = + \\ - \div + = - \end{array}$$

Thus

$$\begin{array}{rcl} +6 \times +2 & = & +12 \\ +6 \times -2 & = & -12 \\ -6 \times +2 & = & -12 \\ -6 \times -2 & = & +12 \\ +6 \div +2 & = & +3 \\ +6 \div -2 & = & -3 \\ -6 \div +2 & = & -3 \\ -6 \div -2 & = & +3 \end{array}$$

Ratio and Proportion

Ratio is another name for division. Thus the ratio 2 to 4 is the same as 2 divided by 4. When the numbers 2 and 4 are written as a ratio, the form used is 2:4; when written as division, the form $2 \div 4$ is used. Both have exactly the same meaning. Since division can always be expressed also as a fraction we often see the ratio of two numbers written as a fraction. The ratio of 2 to 4 can be written in the following three ways, all with equal values: 2:4; 2/4; and 2 ÷ 4.

Whether a ratio is expressed as a fraction or in the form 2:4, it is read "2 is to 4" or "the ratio of 2 to 4."

Figure 8–1 shows two pulleys, M and N, and a connecting belt. If the pulley M revolves at a speed of 125 revolutions per minute and the pulley N, to which it is connected, runs at a speed of 300 revolutions per minute, the ratio of the speed of M to the speed of N is as 125 is to 300 or 125:300, which, as a fraction, can be written 125/300 = 5/12.

The words "ratio" and "proportion" are almost invariably used together. A proportion shows that two ratios are equal, as in 1:2 = 2:4 (or 1:2 : : 2:4). This is read either as "1 is to 2 as 2 is to 4" or "the ratio of 1 to 2 equals the ratio of 2 to 4."

The use of proportion may be illustrated by comparing the costs of two kinds of paper. If No. 122 paper costs 12 cents and No. 123 paper costs 23 cents, we may say that the cost of No. 122 is to the cost of No. 123 as 12 is to 23. By ratio and proportion this can be expressed as

Cost of No. 122:Cost of No. 123 : : 12¢:23¢

This can be written in the fractional form as follows:

$$\frac{\text{Cost of No. 122}}{\text{Cost of No. 123}} = \frac{12¢}{23¢}$$

The quantities in a proportion are called terms; they are numbered from left to right as follows:

First		Second		Third		Fourth
10	:	12	: :	5	:	6

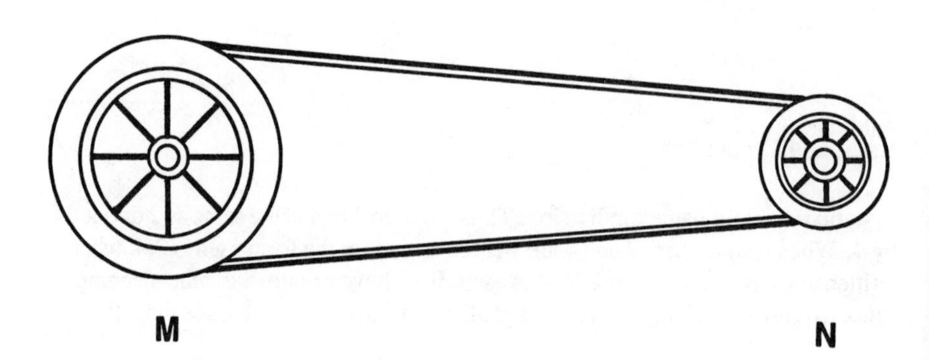

The first and fourth terms are called extremes, the second and third terms are called means (in the sense of middle). Thus, in the proportion 10:12 : : 5:6, the numbers 10 and 6 are the extremes and 12 and 5 are the means.

Work in proportion is simple if one memorizes and learns to apply these three important principles:

1. The product of the extremes is equal to the product of the means.
2. The product of the extremes divided by either mean gives the other mean.
3. The product of the means divided by either extreme gives the other extreme.

In the proportion 36:18 : : 12:x, according to rule 3, the missing extreme,

$x = \frac{18 \times 12}{36} = 6$. In the proportion 9:x : : 6:24 (see rule 2) the missing term,

$x = \frac{9 \times 24}{6} = 36$.

Areas, Surfaces, and Volumes

Plane (flat) surfaces are so common about us that we solve for areas without thinking of geometry at all. When the enclosing lines are straight, it takes at least three straight lines to enclose a surface.

Area of a Square or a Rectangle

The area of a rectangle is equal to the product of its length multiplied by its width. If two sides of a rectangle are 10 inches and the other sides are 5 inches, the area is 50 square inches. Length and width must both be expressed in the same denomination, as, for example, in feet or in inches

Area of Parallelogram

The area of a parallelogram is found by multiplying the base by the altitude. AC is the altitude and BD is the base, so the area of the surface is 4 × 10 or 40 square inches. The possibility of error here is that of multiplying the length AB by the length of BD. This is incorrect because AB is not an altitude. The altitude is the perpendicular distance between two parallel sides.

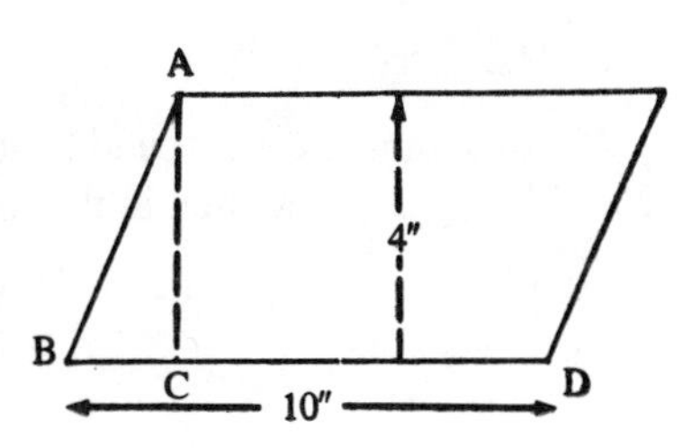

Area of a Triangle

The area of a triangle may be found in several ways, according to the kind of triangle that one is considering and according to the values that are given. The area of a triangle is equal to the product of one half the base by the altitude. The altitude of a triangle is always the perpendicular distance from a vertex (any of the points where the sides of a triangle intersect is a vertex) to the opposite side, called the base.

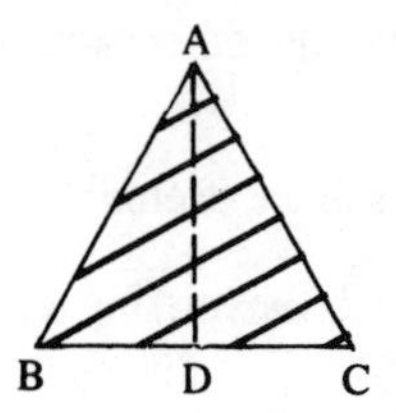

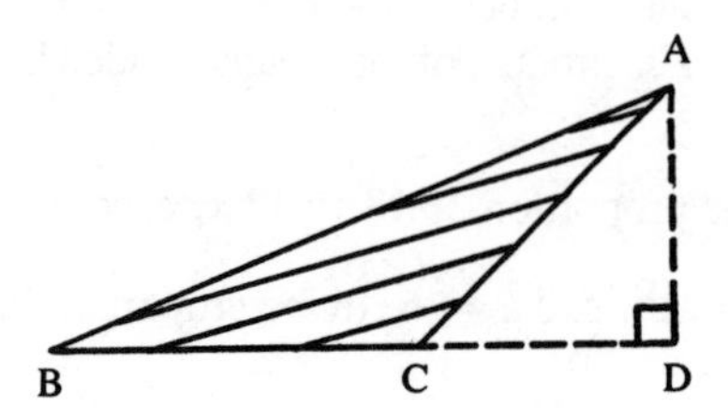

This means that in either of the above figures the area (shaded) is $\frac{1}{2} \times (AD \times BC)$. If BC is 5 inches long and AD is 6 inches long, the area is 15 square inches, $\frac{1}{2} \times (5 \times 6) = 15$. Note particularly that the altitude of a triangle may be outside of the triangle itself, as shown in the second triangle above. The area of an equilateral triangle (three equal sides) is $\frac{1}{4}(\text{side})^2 \div 3$. The area of a triangle in general is

$$\text{Area} = \sqrt{S(S-a)(S-b)(S-c)}$$

$S = \frac{1}{2}$ (Sum of the sides) $= \frac{1}{2}(a + b + c)$, a,b,c,—three sides.

Area of a Polygon

A surface that is enclosed by three or more straight lines is called a polygon. This includes triangles, squares, hexagons, pentagons, etc. The area of a polygon is no. of triangles × area of one regular triangle in the polygon.

Area of a Circle

The area of a circle is found by multiplying the square of the radius by 3.14 or 22/7. This may be written in the form of

$$\text{Area} = \pi r^2 \text{ or } \frac{\pi D^2}{4}$$

where r and D represent the radius and diameter of the circle, respectively.

Area of the Surface of a Cylinder

To find the area of a cylindrical surface, find the distance around the circle and multiply this distance by the height. If the diameter of a tank is 20 inches and the height 5 feet, the surface area of the cylinder may be found by the formula:

$$\begin{aligned} \text{Area} &= 3.1416 \times \text{diameter} \times \text{height} \\ &= 3.1416 \times 20 \times 60 \\ &= 3769.92 \text{ square inches} \end{aligned}$$

Volume of a Cube

The volume of a solid with a rectangular cross-section may be found by multiplying the area of the base by the height or the product of its length, width, and height. The volume of a cube may be found by the same method as for any other solid with a rectangular cross-section. However, as the edges of a cube are all equal, the product of its length, width, and height is the same as the cube of one of its edges. Therefore, we say the volume of a cube is equal to its edge cubed.

Volume of Cylinders

A cylinder is similar to a solid with a rectangular cross-section. The surface of a cylinder, not including the base, is a rectangle. The total volume of a cylinder is equal to the area of the base times height. If the base is a circle, the volume may be expressed as:

$$\begin{aligned} \text{Volume} &= 3.1416 \times r^2 \times \text{altitude} \\ r &= \text{radius of the base} \\ \text{altitude} &= \text{height} \end{aligned}$$

PROFESSIONAL AND "REAL WORLD" MATHEMATICS COMPARED

The mathematics of the "real world" has certain characteristics that distinguish it from that used by people in their professional work.

1. Mathematics of the real world uses more estimation (benchmark, decomposition-recomposition) than actual calculations.
2. Even the ordinary calculations of the "real world" have many more factors to contend with. For example: Given a 10 inch long piece of wood, show how many pieces of 2 inch width can be cut from this piece. (Answer: 5 pieces.) The answer given is a mathematical answer. In the "real world" the answer will not be 5 or the five pieces will not have 2 inches as their

width as there is wastage when the pieces are cut. This must be kept in mind or the results could cause unexpected problems.

3. The optimal solution of a problem in the theoretical situation may not be the optimal solution of the corresponding practical problem. An individual may sacrifice a payoff for practical reasons.
4. Mathematics in the academic setting is a cognitive activity, whereas mathematics in the "real world" is a combination of cognitive, affective and psychomotor activities involving real feelings, emotions, and psychic aspects.

MATHEMATICAL CONCEPTS FOR THE EVERYDAY WORLD

More and more science and technology are permeating our society and with this the need for more mathematics to understand the scientific and technological concepts. The new level of mathematical competencies, skills, and attitudes toward mathematics required of modern citizenry is much higher than what was expected 25 years ago. The following represents the skills and competencies considered necessary for adults to participate effectively in contemporary society:

1. Numbers and Numerals
   - Express a rational number using a decimal notation.
   - List the first ten multiples of 2 through 12.
   - Use the whole numbers (four basic operations) in problem solving.
   - Recognize the digit, its place value, and the number represented through billions.
   - Describe a given positive rational number using decimal, percent, or fractional notation.
   - Convert to Roman numerals from decimal numerals and conversely (e.g., data translation).
   - Represent very large and very small numbers using scientific notation.
2. Operations and Properties
   - Write equivalent fractions for given fractions such as ½, ⅔, ¾, and ⅞.
   - Use the standard algorithms for the operations of arithmetic of positive rational numbers.
   - Solve addition, subtraction, multiplication, and division problems involving fractions.
   - Solve problems involving percent.
   - Perform arithmetic operations with measures.

- Estimate results.
- Judge the reasonableness of answers to computational problems.

3. Mathematical Sentences
   - Construct a mathematical sentence from a given verbal problem.
   - Solve simple equations.

4. Geometry and Measurement
   - Recognize horizontal lines, vertical lines, parallel lines, perpendicular lines, and intersecting lines.
   - Recognize different shapes.
   - Compute areas, surfaces, volumes, densities.
   - Understand similarities and congruence.
   - Use measurement devices.

5. Relations and Functions
   - Interpret information from a graphical representation.
   - Understand and apply ratio and proportion.
   - Construct scales.

6. Probability and Statistics
   - Determine mean, average, mode, median.
   - Understand simple probability.

7. Mathematical Reasoning
   - Produce counter examples to test invalidity of a statement.
   - Detect and describe flaws and fallacies in advertising and propaganda where statistical data and inferences are employed.
   - Gather and present data to support an inference or argument.

8. General Skills
   - Maintain personal bank records.
   - Plan a budget and keep personal records.
   - Apply simple interest formula to calculate interest.
   - Estimate the real cost of an item.
   - Compute taxes and investment returns.
   - Appraise insurance and retirement benefits

There are thousands of people from many walks of life who look back to their mathematics lessons with regret, either because they never really understood the point of them all, because the pleasure of them ceased when they left school, or because much of what they once learned would be useful if only it could be recalled. Special approaches to the teaching of mathematics are needed if we as educators want to help the citizens of this country to become familiar with, to apply, and to feel comfortable with the minimal mathematics they need to function effectively and without fear of failure in every aspect of living in today's world.

Chapter 9

Selection, Adaptation, and Development of Curricula and Instructional Materials

John F. Cawley

Selection is the process of choosing. *Adaptation* is the process of changing. *Development* is the process of creating and producing. Fundamental to each of these is implementation, for without effective plans to implement the curricula or the materials to which we have devoted energy and expertise our efforts are in vain.

GENERAL CONSIDERATIONS

Decisions regarding selection, adaptation, and development of teaching materials are difficult ones for a number of reasons. Among these are:

- varying opinions as to what should be taught to the learner
- different levels and areas of knowledge among the professional staff who make decisions
- different understandings of the characteristics and needs of the learners
- variations in instructional skills among staff
- different teaching and management styles
- varying levels of material awareness
- different degrees of emphasis on skills, concepts, and problem solving

No single factor should serve as the only standard or criterion to guide selection, adaptation, or development. Tradeoffs are necessary and a balance among criteria is essential. Three factors, however, seem paramount. The first of these relates to needs and the question as to whose needs, the system's, the teachers', or the learners', is to be given priority. The second factor is completeness. That is, any effort should be comprehensive and meet the needs of a range of learners for an extended period of time (e.g., K–6, K–12, 7–12). Third, development at the local or state level is a more complex process than selection or adaptation. Development should be considered only when there is a need for unique materials and no other means of obtaining what is required.

Opinions of What To Teach

The processes of selection, adaptation, and development are guided by the decision as to what is to be taught. Some might argue that learning-disabled children should be provided a comprehensive and total approach to mathematics. Others might argue that specific skills (i.e., computation) should be the focus of the program. Still another group might suggest that the program focus on assessed needs. Some may wish to focus on the primary grades and others may desire to focus upon the secondary level. As we have seen, there is no avoiding this question of content. Needless to say, a general consensus goes a long way.

Levels and Areas of Knowledge

The staff of any single school district differs in its levels ana areas of knowledge in mathematics and special education. It is not uncommon to find secondary school mathematics teachers who plead innocent when it comes to knowledge of the mathematics of primary grades. Primary teachers are often in similar situations relative to secondary school mathematics. Special education teachers may or may not have a background in mathematics. These variations clearly suggest the need for a cooperative approach to selection, adaptation, and development.

Learner Characteristics

As we have noted, it is also difficult to obtain a generalized agreement as to what constitutes a learning disability. It is even more difficult to specify the characteristics of each and every child and to obtain reliable agreement as to the needs of the individual. Whatever the characteristics of the learning-disabled child, they must be interpreted in language and terms that are understood by all those who serve the child. Descriptions such as "He is at the third-grade level in addition" or "She is in the fifth grade and functioning two years below expectancy" are relatively meaningless statements. They do not describe the characteristics of the learner, nor do they assist in the development of program decisions. At the very least, program decisions need to be arrived at with high degrees of consistency. That is, two or more teachers looking at the same information should come to similar conclusions as to what should be done. Examine the following examples:

> *Report A:* Jerry is a fourth grade youngster who is functioning at the sixth grade level in mathematics.

or

Report A: Jerry is a fourth grade youngster who is able to do computation with whole numbers through three-digit divisors and multiplication and division of fractions and mixed numbers. Difficulty is constantly encountered in the addition and subtraction of fractions when the denominators differ. Jerry identified and named circle, square, trapezoid, and rhombus as two-dimensional shapes, and a cube, sphere, and cylinder as three-dimensional shapes. Jerry missed all test items on the topics of *motion*.

Report B: Carol is a fourth grade youngster who is functioning at the second grade level in mathematics.

or

Report B: Carol is a fourth grade youngster who counts to 45 by 1s and does skip counting with 2s to 24. Carol does not count by 5 or any other number. Carol performs single-digit addition and subtraction. Carol has excellent concepts of congruence, similarity, and symmetry. Carol also has excellent concepts of discrete parts, blended parts, and replaceable or interchangeable parts.

The descriptors used in each instance carry different messages to different people. The first description of Jerry fails to indicate relative weakness (e.g., his problem with differing denominators) or strengths (e.g., his knowledge of shapes and figures). The first description of Jerry suggests a youngster who is functioning well above expectancy and without specific needs. The second descriptor informs us that Jerry has both strengths and weaknesses in fractions. Before making any decision about his lack of proficiency in motions, one would have to check his cumulative record and determine if he has had exposure to the topic.

In all likelihood Carol has been exposed to a comprehensive mathematics curriculum. Concept capability in geometry and fractions is evident. Carol's concepts of numbers are undetermined, although levels of computation are stipulated.

What program decisions should be made? Would you agree with the following?

Jerry: Assistance with the development and understanding of equivalence classes, congruent part of whole and interchangeable parts; move to addition and subtraction of fractions; teach concepts of slides, flips, and turns.

Carol: Continue efforts in concepts of geometry and fractions. Avoid computation with latter. Determine meaningfulness of counting. If satisfactory, go to concepts of addition and subtraction (e.g., reversibility, nonreversibility, and zero). Develop concepts of regrouping (not renaming) with objects.

Knowledge of Instructional Adjustments

Professional staff members vary in their knowledge of instructional techniques. Some individuals know numerous techniques, whereas others know only a few. Instructional adjustments consist of changes in the manner of presentation (e.g., from paper and pencil to manipulatives), modifications and changes in algorithms, and adjustments in content, level, or sequence.

Adjustments in manner of presentation are required to meet the needs of individual learners and to represent mathematics principles and concepts. It may be that a child is a poor reader and not able to read problems of a level at which he or she can do problems if they are presented in another manner (Ballew & Cunningham, 1982). It may be that a youngster does not understand the derivation of $\frac{1}{9}$ in the example $\frac{1}{3} \times \frac{1}{3}$ when written or verbal explanations are given. The teacher may elect to represent this item with pictures. It may be that a child has habituated an inappropriate algorithm and that a change from paper and pencil to manipulatives can be used as an intervention.

Modification in algorithms may be used to assist individuals to develop appropriate concepts and to generalize appropriate responses until a conventional algorithm is internalized. For example, two children performed as shown below:

A	B
6344 5478 11 111 2 3627 14 ——— 15459	6344 5478 3627 ——— 14 13 14 19 ——— 15459

Each of these youngsters utilized a different algorithm. Each has the same incorrect answer. Each used the algorithm correctly. Each made a simple computational error.

No change is needed in the algorithm. Attention should be directed to the carelessness in computation. Examine the illustrations that follow:

A	B
6344 −3627 ——— 1727	6344 −3627 ——— 3727

Each of the youngsters used an unacceptable algorithm. The algorithm used by A is unacceptable and incorrect. The algorithm used by B is unacceptable as far as B has gone. B had an idea. What B was supposed to do was "add one" when the top number was fewer than the bottom number. B did this. What B failed to do was compensate for this by adding 1 to the 2 to make it 3 and 1 to the 3 to make it 4, giving:

$$\begin{array}{r} 6344 \\ \overset{4}{\cancel{3}}6\overset{3}{\cancel{2}}7 \\ \hline 2717 \end{array}$$

The teacher decided to eliminate the traditional orientation and went to the following:

$$\begin{array}{r} 63 \\ -7 \\ \hline \\ 59 \\ -7 \\ \hline 52 \\ +4 \\ \hline 56 \end{array} \qquad \begin{array}{r} 634 \\ -367 \\ \hline \\ 599 \\ -367 \\ \hline 232 \\ +35 \\ \hline 267 \end{array} \qquad \begin{array}{r} 6344 \\ -3672 \\ \hline \\ 5999 \\ -3672 \\ \hline 2327 \\ +345 \\ \hline 2672 \end{array}$$

The alteration in the use of algorithms is designed to place the youngster at the introductory level. Previous strategies do not work and thus do not interfere with the newer approach. Ultimately, the individual will be guided to the conventional algorithm.

Another instructional adjustment involves curriculum content, curriculum level, or curriculum sequencing. These adjustments are made by the teacher who wishes to present another topic or to develop more extensive background for the child. In order to do so, the teacher may find it necessary to adapt instructional materials and to restructure the class. The inclusion of alternative content is related to the question "What is to be taught." However, it is not simply a switch from one topic to another (e.g., division with whole numbers to division with decimals) because such a switch may also require that the level of presentation be modified (e.g., $372\overline{)4863}$ to $3.2\overline{)3.2}$) to ensure proper understanding. It may also be necessary to sequence the move from $372\overline{)4863}$ to $3.2\overline{)3.2}$ by going through $.3 \times 1 = ?; .3 \times .1 = ?; 3.2 \times .1 = ?; 3\overline{)3}; 3\overline{)3.3}; 3\overline{)3.2}; .3\overline{)3}; .3\overline{)3.2}; 3.2\overline{)3.2}$ and within this explain why the decimal point is moved.

This is where "tradeoffs" have to be considered. If one wishes to devote more time to multiplication, that time has to be taken from something else. If one wishes to provide extra pages for multiplication, then these pages must take the place of something else.

Material Awareness

Some teachers are collectors. No matter what the topic these individuals have shelves and closets full of materials. Others manage with a single text and workbook. Still others are somewhere in between. Material awareness requires that a teacher be familiar with the full range of special education and regular education materials. This is a difficult task because producers tend to disseminate to select mailing lists or potential customers. Accordingly, there needs to be considerable sharing of catalogs, material listings, and advertisements among all members of the staff.

Preferences also influence our knowledge of materials. One who is solely interested in computation will not collect material related to problem solving or other topics. One who teaches at the secondary level is not likely to have a great deal of interest in elementary materials.

An extensive assessment of media and materials for the handicapped has been completed by Educational Testing Service (Vale, 1980). This effort analyzed the needs of 28,044 respondents. Among the more relevant findings are the following:

- Priorities for children aged 11 and under tend to be in the basic skills areas whereas those for students 12 or older are marked by a job preparation/vocational education cluster.
- Teachers' primary reason for needing materials development, cited across all handicap categories, is lack of variety in available materials.
- Major inadequacies with respect to materials content are the failure to provide sufficient repetition or reinforcement of the material and inappropriateness with regard to interest, concept, and language/vocabulary levels and to the child's life experiences.
- High among the inadequacies of materials are inability to be used independently by the student, needs too much supplementation by the teacher, not adaptable for a variety of disabilities, doesn't suggest alternative learning strategies, doesn't provide for different entry levels, and doesn't provide for evaluation of student progress.
- About 82 percent of all teachers surveyed indicated a great or moderate need for additional resources, time, and training to develop their own materials.

- More teachers and supervisors requested information about available instructional materials, conferences and workshops, and newly developed materials and adaptive devices than any other topics.
- Even for the most frequently requested topics, more than a third of the teachers and almost as many supervisors report that such information is not readily accessible. The most glaring example of this information gap is the topic of newly developed materials.
- The major reasons for seeking information are matching specific materials to individual learning needs, curriculum planning, and identification of materials for preview.
- The least accessible information sources to both teachers and supervisors are information retrieval systems.

Sources of materials are limited. The listing of materials in the *Master Index to Special Education Materials* (University of Southern California, Los Angeles) is one important source. This index, as extensive as it may be, is not all inclusive. For example, the listings under geometry focus on shapes and figures. No reference is made to topological concepts such as order-constancy or betweenness or to relations such as motions and parallelism.

The catalogs of various publishers describe many materials. In addition, they also contain excellent suggestions and ideas for teaching or the developing of instructional materials. One should be on as many mailing lists as possible.

The *Arithmetic Teacher*, a journal of the National Council of Teachers of Mathematics, is a must for every special educator. Each issue contains meaningful and practical suggestions on how to teach specific concepts or how to prepare materials. Over the years one can keep an inventory of specific topics and develop outstanding recommendations for materials or teaching.

SELECTION

Selection has been defined as the act of choosing. How does one person or a system go about making choices? To begin with, we make choices on a basis of need, but whose need? If, for example, it is decided that the primary consideration is the needs of the child, all other factors are secondary. If, however, a decision is made to meet the needs of the staff, or the system, or the subject area, or the budget, the needs of the child must be interrelated. Let us illustrate the selection process by examining the move to microcomputers in many school districts. The following questions are relevant:

- Is there evidence that suggests that the needs of learning-disabled children can be effectively met in mathematics with microcomputers?

- Are systems moving toward the use of microcomputers because they feel they will be out of date if they don't? Are they simply trying to keep up with the Joneses?
- Do the programs provide for the youngsters in a manner in which teachers or other materials do not?
- Will the child be the beneficiary of a comprehensive program of mathematics? If not, how will other facets of the program be presented?
- Do learning-disabled children develop proficiency in computation with whole numbers more effectively with computer-aided instruction than they do with teachers? Do they come from farther behind at a faster rate to catch up more quickly? Do they understand what they are doing?
- What will be the long-term effect? That is, what about the child whose program is presented by microcomputer over a period of five to ten years?
- How many computers are needed? If a group of five youngsters requires ten minutes each of computer-aided instruction, what do the other four do for the 40 minutes when the terminal can be used by only one individual?
- What is the relative quality of the content and the manner in which it is presented? Is the software something special? Does it do something for the teacher and child that is not done by a paper and pencil program?

Most software is quite traditional. Word problem programs rarely capitalize on the diagnostic capabilities of the hardware. Exhibit 9–1 illustrates one diagnostic format.

Lists of questions similar to those about microcomputers and computer-assisted instruction are essential in the process of selection. They represent one response to the appropriate set of questions that substantiates the need.

The primary question for the special educator is to determine what will be taught. Earlier, in Chapter 2, Blankenship cited a perspective from mathematics suggesting that the problem of what to teach was not as difficult as the problem of how to teach. While that may be true of general education, it does not seem to be the case for special education. Special education has concentrated on how to teach for at least a decade and there is a variety of instructional alternatives available. The more crucial issue seems to be what to teach, when to teach it, and in what sequence is it best taught. Special educators are prone to identify arithmetic computation as the set of mathematics for the learning disabled. Only scant attention has been paid to the selection of other content.

Content

There are two possible approaches to content selection. One is to start with some representation of the universal set of mathematics. The other is to start with an empty set and construct the set of content. The first alternative is recommended,

Exhibit 9–1 Diagnostic Format for Word Problems

A tall boy had 3 apples left after he gave 4 apples and 6 pears to a friend. How many apples did the tall boy have to start with?

- If 7, say, "Good job. You did not get fooled by thinking that the word *left* meant to subtract. Also, you did not get fooled by adding in the pears. The question only asked about apples. You did a good job of thinking."
- If 13, ask, "Why did you add all the numbers? The question only asked about apples. The question did not ask about the pears. Read the problem and try again."
- If 9, ask "Why did you add the 3 and the 6? The question asked you about apples, not pears. Look the problem over and try again."
- If 10, ask, "Why did you add the 4 and the 6? The question asked you only about apples, not pears. Look the problem over and try again."
- If 3, ask, "Why did you subtract the 3 from the 6? Does the word *left* make you think about subtraction? In this problem, the word *left* should not tell you to subtract. Read the problem and think carefully. Be certain you know what the question asks."
- If 1, ask, "Why did you subtract? Did the word *left* make you think about subtraction? Read the problem again. Think about the problem. Don't get fooled by a word such as *left*. Try again."
- If 2, ask, "Why did you subtract 4 from 6? The problem asks only about apples. You subtracted apples from pears. Check carefully before you decide to subtract. Now, try again."

because it enables one to review all topics, to eliminate those determined to be of lesser importance, and to prioritize the remainder by level and sequence. This approach might be initiated with the purchase of the *Content Authority List* (Pennsylvania Department of Education), which is a listing of the topics of mathematics covered in the elementary grades. A teacher or a team could scrutinize the list, rate items as to their importance, and negotiate discrepancies. A second approach would be for the teacher or the team to obtain a list of all topics covered in the pre-K through secondary program of the local education agency.

If the youngster is enrolled in a regular class the content needs to be representative of the regular class. The child cannot be sent to the resource room during a time when measurement topics are being covered in the regular class to receive assistance in subtraction. The child with problems in mathematics needs more mathematics, not less. If the youngster is enrolled in a self-contained special education unit, and it appears that placement is long-term, the self-contained classroom needs to have a comprehensive curriculum in place. If enrollment is viewed as short-term (e.g., two years) the program needs to fill in the gaps in the background of the child and prepare her or him for reentry to the regular classroom.

Selection must also involve a consideration of primary and secondary content. Primary content would consist of material to be presented in the dominant placement. If the dominant placement is the regular classroom, content selection must conform to this placement. Secondary content would consist of material that is complementary to or supplementary to the dominant content. This could include expansion or intensification on specific topics. It might include an emphasis on problem solving or applications. The regular classroom may be highly concept oriented and the supplemental content could focus more upon skills.

Level

A second selection consideration has to do with level. Once it has been decided what will be taught, it is necessary to determine when the youngster is ready for this content and when is the preferred time to present it. The present level of functioning in mathematics, cognitive maturation, previous experiences, rate of progress, and task requirements all enter into the picture. For example, it is common to find computation in fractions introduced into the program at about age nine or grade 4. For some, this may be their first formal experience with fractions. The true need might be conceptual and not computational. For others, numerous concepts and principles of fractions may have been part of their program since kindergarten. This latter group might be more ready for computation than the first group.

Sequence

Sequencing is a third consideration. Does the system or staff have a preferred sequence? Although it is traditional to introduce fractions in a sequence from addition to subtraction to multiplication and to division, it may be that the move to multiplication and division is simpler and could be undertaken prior to addition and subtraction.

Materials

In tandem with the selection of content is the selection of curriculum and instructional materials. The following list by Gall (1981) identifies features of curriculum materials that may facilitate learning and, therefore, could be used to provide the selection process.*

---

*Gall, M.D. (1981). *Handbook for evaluating and selecting curriculum materials*. New York: Allyn and Bacon. Copyright © 1981, used with permission.

1. Advance Organizer: A preview, usually in the form of an introductory paragraph, that gives the learner a conceptual framework for assimilating the learning experience to follow.
2. Clarity: The clear presentation, to the learner, of vocabulary, syntax, concepts, examples, and content organization.
3. Cueing: Items pointing to information and ideas important for the learner to understand and remember. Cueing can be in the form of headings, margin notes, use of contrasting typeface, or explicit statements telling the learner that a particular fact or idea is important.
4. Diagnostic Tests: An assessment procedure for indicating skill areas in which the student has strengths or weaknesses. Diagnostic tests provide the basis for individualized instruction.
5. Enthusiasm: A communication style that conveys the excitement of a particular field of knowledge to the learner. Tone of voice (e.g., in an instructional film) and personalized writing style are techniques for projecting enthusiasm.
6. Examples: Particular facts, instances, or aspects to illustrate a more general concept or principle.
7. Feedback: Procedures for providing information to learners on the quality or accuracy of their performance in practicing new knowledge and skills.
8. Glossary: A list of unfamiliar and technical terms presented in the curriculum materials so that the learner can focus attention on them.
9. Independence Level: Inclusion of instructional experiences appropriate for the learner's level of independence.
10. Inserted Questions: Placement of questions before, in, or after an instructional message to help the learner engage in "active" reading, viewing, or listening.
11. Manipulatives and Realia: Objects or activities used to relate classroom teaching to real life.
12. Models: Descriptions or analogies used to help the learner visualize something that cannot be directly observed.
13. Objectives: Explicit statement of an instructional unit's goals so that the learner knows beforehand what knowledge and skills s/he needs to master.
14. Outline: A summary of the main ideas and their sequence in an instructional unit, the purpose of which is to cue the learner to these ideas and to provide a framework for organizing them.
15. Perceived Purpose: Presentation to the learner of reasons for studying a particular unit of instructional content.

16. Practice: Opportunity for the learner to rehearse the knowledge or skills contained in an instructional unit.
17. Prerequisites: Explicit statement of the knowledge and skills needed by the learner if s/he is to profit from a particular instructional experience.
18. Recycling: Opportunity for the learner to restudy an instructional unit if s/he fails a corresponding test.
19. Reinforcers: Provision of an extrinsic reward (e.g., tokens, praise, grades, a pleasurable activity) to the learner upon successful completion of an instructional experience.
20. Self-Check Test: An assessment device that enables the learner to check his or her level of mastery. Results of the self-check test are not used as a formal evaluation of the learner's performance by the teacher.
21. Variety: Several different instructional strategies, formats, and media used to convey instructional content.

ADAPTATION

Adaptation is the process of changing. Specific to mathematics and the learning disabled, there are many factors to consider relative to change. These might be grouped under the following categories:

1. curriculum—content, levels, and sequences
2. instruction—manner
3. material

Lambie (1980) suggests that any approach to the changing of materials, instruction, and assignments should be systematic. She describes three types of changes: adaptation of materials, modification of instruction, and alteration of assignments. As guidelines for making these changes, Lambie suggests the following:

1. Use the change process only when there is a mismatch between learner and material.
2. Keep changes simple.
3. Confirm mismatch by evaluating changes made.
4. Minimize teacher time in making changes.
5. Keep combinations consistent.
6. Know the strengths and weaknesses of the material.
7. Know the strengths and weaknesses of the learner.

The use of the above list of principles is applicable, according to Lambie (1980), in dealing with the following questions:

1. What if there are too many items, questions, etc.?
2. What if there is not enough repetition?
3. What if a lack of feedback results in problems when students use the material independently?
4. What if the visual presentation is too confusing?
5. What if the students do not remember or understand the directions?
6. What if the material, lesson, or assignment is not interesting?
7. What if the product is not durable?
8. What if the material/lesson moves too rapidly?
9. What if the lesson is too complex?
10. What if the presentation sequence of skills/concepts is too brief or choppy?
11. What if significant information is not focused upon?
12. What if the language level is too high/different?
13. What if purchased materials assess recall only?
14. What if the material/lesson is biased?
15. What if the verbal response is a problem?
16. What if the motor response is too difficult?

The primary purpose of adaptation is to meet individual needs. This is not a new problem or one that is unique to the learning disabled. Flournoy (1960) lists a set of variations that teachers judged to be essential to meet individual needs in the regular classroom.

Examples of content variations are:

- adding topics for the fast learner that are not ordinarily found in the course of study, such as finding median and mode and learning to count with a base other than 10, while perhaps omitting a few infrequently used topics for the slow learner
- varying the level of difficulty undertaken on any one topic, for example, encouraging faster learners to master the 10s, 11s, and 12s in multiplication and division or having the slower worker do exercises in dividing decimal fractions that involve only whole numbers and tenths as a divisor rather than spending a great deal of time trying to master division by hundreds and thousandths
- varying the content of practice exercises
- providing rapid learners with more difficult horizontal enrichment involving problem situations in which research is necessary to gather data, such as finding the cost of advertising in a local paper

- providing rapid learners with more difficult types of horizontal enrichment that stimulate the mathematical appetite and foster an interest in mathematics as a hobby
- allowing fast learners to study certain selected topics normally taught in a higher grade, for example, progressing faster with the learning of addition, subtraction, multiplication, and division facts; studying the meaning of percent in the fourth or fifth grade; and tackling areas of parallelograms and circles in the sixth grade. At the same time a whole school plan for delaying the teaching of a few selected topics for the slower learners might be inaugurated, such as delaying the counting by groups of 5s, 2s, and 3s for a grade or more and delaying multiplication and division of decimal fractions until after grade 6.

Examples of varying teaching methods and materials are:

- follow-up reteaching of new skill to slower learners
- frequent review of steps in a process for slow learners
- more closely teacher-directed reading of textbook for slow learners
- longer and more frequent use of concrete materials with slow learners
- more independent use of textbooks by fast learners
- more mental arithmetic exercises for fast learners
- use of encyclopedia and other such materials by fast learners to investigate certain arithmetic topics
- differentiated test items with more difficult, unusual, or challenging items for faster learners

Adaptation of curriculum either by adding or dropping topics or by changing levels or sequences takes place continuously. The experiences, knowledge, and success of staff, the gains of children, and evaluations guide the process. New topics become priorities; existing priorities change. For example, it is unlikely that one could find curricula for the learning disabled with a focus on nominal numbers, yet, the requirements of the future will unquestionably demand more attention to this topic.

Instructional Adaptations

Instructional adaptations are generally the most efficient for the knowledgeable and competent person. These range from an overall commitment to a specific model (e.g., Data-Based Instruction (Blankenship, Chapter 2 above) to more generalized orientations. Instructional adaptations are efficient because they can be accomplished with relatively little preparation and without extensive material adaptation. This section will discuss instructional adaptation as a change in the manner in which mathematics principles and concepts are represented for the learner within the framework of a specific model, as illustrated in Table 9–1. This model is entitled the Interactive Unit (IU).

The IU is composed of 16 cells that format the interactions between teachers and learners or among teacher, learner, and material. The IU provides a means for developing equivalent representations of skills and concepts in mathematics. It is not hierarchical, that is, in some instances a concept may be qualitatively represented manipulatively whereas the use of the written symbols may be by rote. (A good illustration of this is the task $\frac{1}{2} \times \frac{1}{3} = \frac{1}{6}$ when the child asks for an explanation as to the meaning of the "$\frac{1}{6}$." More learning-disabled children know how to do $\frac{1}{2} \times \frac{1}{3}$ in its written form than with cubes or sticks.)

The IU does not seek preferred modalities. That is, it is not used to describe children as visual learners or auditory learners. It is a means of developing alternative modality usage for the purposes of representing concepts and skills in different ways and of partialing out the effect of one disability (e.g., written symbolic difficulties) on a performance area (e.g., word problems).

The IU provides a basis for systematic variation in the manner of presentation and for the systematic adaptation or development of instructional materials. Moves throughout the matrix are orderly and deliberate. Note that any move across the row results in a modification of teacher behavior, whereas a move up or down a column results in a change in learner behavior. The IU provides the teacher with a means of (1) reteaching or reevaluating a skill or concept using different interactions, (2) maintaining an awareness of the need for his or her presence during the instructional or follow-up activity, and (3) developing alternative formats for instructional materials.

The use of selected interactive combinations enables the teacher to provide varying levels of abstractness in the representation of specific concepts or skills. Frequently, when a child is experiencing difficulty and the teacher elects an instructional alternative, the alternative does not represent the original algorithm with which the difficulty was encountered. To illustrate, assume a child was experiencing difficulty with $12\overline{)2448}$ and the teacher wanted to rework the item with manipulatives. In many instances a change would take place and the child would be introduced to an addition $\begin{array}{r}12\\+12\\\hline 24\end{array}$ or subtraction $\begin{array}{r}2448\\-\quad 12\\\hline 2436\end{array}$ algorithm. The next step would be to move back to $12\overline{)2448}$ and assume that the previous illustrations via addition or subtraction enable the learner to proceed. However, the teacher is confusing the concept of division, determination of equal units, with skill in the use of the algorithm. Following the format of the IU, one would not change strategy (i.e., move from $12\overline{)2448}$ to $12 + 12$) but maintain the same algorithm and represent it differently, as illustrated in Figure 9–1.

Table 9–1 The Interactive Unit (IU)

LEARNER – OUTPUT \ TEACHER – INPUT	MANIPULATE (M)	DISPLAY (D)	SAY (S)	WRITE (W)
MANIPULATE (M)	TEACHER MANIPULATES LEARNER MANIPULATES	TEACHER DISPLAYS LEARNER MANIPULATES	TEACHER SAYS LEARNER MANIPULATES	TEACHER WRITES LEARNER MANIPULATES
IDENTIFY (I)	TEACHER MANIPULATES LEARNER IDENTIFIES	TEACHER DISPLAYS LEARNER IDENTIFYS	TEACHER SAYS LEARNER IDENTIFIES	TEACHER WRITES LEARNER IDENTIFIES
SAY (S)	TEACHER MANIPULATES LEARNER SAYS	TEACHER DISPLAYS LEARNER SAYS	TEACHER SAYS LEARNER SAYS	TEACHER WRITES LEARNER SAYS
WRITE (W)	TEACHER MANIPULATES LEARNER WRITES	TEACHER DISPLAYS LEARNER WRITES	TEACHER SAYS LEARNER WRITES	TEACHER WRITES LEARNER WRITES

Manipulate (M)—Manipulation of objects—piling, arranging, moving

Display (D) (Teacher input only) Presentation of displays—pictures, arrangements of materials

Say—Oral discussion

Write—Written materials—letters, numerals, words, signs of operation, and the marking of these types of material

Identify (Learner output only) Selection from multiple choices of nonwritten materials (pictures, objects)

Figure 9–1 Instructional Adaptation for Algorithms

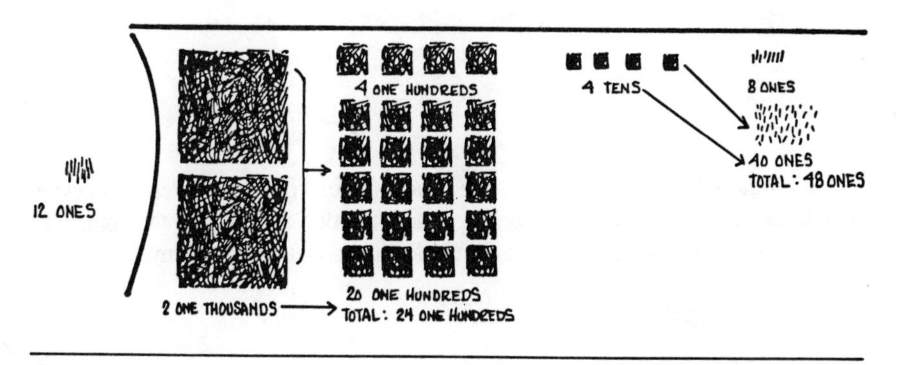

The first task is to have the learner make as many sets with this many in a set (point to divisor) as you can with these (point to representation of 1000s). However, there are only two representations of 1000 and the learner is unable to construct any set with 12 of the 1000 in a set, thus, the source of the 0 in $\begin{array}{r}0\\12\overline{)2448}\end{array}$.

At this point, the instructor may say, "Watch me and convert the 1000s to 100s," and direct the learner to make as many sets with this many (point to divisor) in each set as you can with these (point to 100s). The learner pushes the items together and makes two sets, each with 12 in a set. This initiates the learner's understanding of the source of 02 in $\begin{array}{r}02\\12\overline{)2448}\end{array}$.

The procedure continues until $\begin{array}{r}0204\\12\overline{)2448}\end{array}$ is determined. This same approach can next be done with numerals rather than with manipulatives. It would look like this:

Original: $12\overline{)2448}$

Adaptation:

```
12)1000  1000    100  100  100  100    10  10  10  10    11111111

      0                  2                  0                   4
12)1000  1000    100  100  100  100    10  10  10  10    11111111
            \    100  100  100  100                 \    11111111111111111111
             `-> 100  100  100  100                  `-> 11111111111111111111
                 100  100  100  100
                 100  100  100  100
                 100  100  100  100
```

And finally, the traditional

$$\begin{array}{r} 204 \\ 12\overline{)2448} \\ \underline{24} \\ 48 \\ \underline{48} \end{array}$$

Both of these last two illustrations are conducted in write/write interaction even though they are quite different. The expanded notation accentuates the contribution of place value, something that should be stressed in all programs.

Teacher Participation

The IU provides alternatives for direct or indirect teacher involvement in the instructional activity, as shown in Table 9–2.

Teachers can organize small or large group instruction with the IU. The guiding factor should be whether or not their presence is required. There are a number of alternatives. One is for the teacher to use a common input for all youngsters and to have different responses from different youngsters as follows:

Group 1 (5 students)	Group 2 (9 students)	Group 3 (6 students)	Group 4 (3 students)
"Manipulate"	"Identify"	"Write"	"Say"

The teacher can represent a concept via manipulation and individuals can respond according to the group to which they are assigned. Note that group 4 is small. The reason for this is that aural exchanges must take place in sequence and only one youngster can perform at a time. In all other interactions simultaneous learner performance is possible.

Table 9–2 Teacher Involvement and the Interactive Unit

	M	*D*	*S*	*W*
M	TR	NTR	TR	NTR
I	TR	NTR	TR	NTR
S	TR	TR	TR	TR
W	TR	NTR	TR	NTR

TR = Teacher presence required

NTR = No teacher presence required

The IU is also of use in organizing classes into groups. It not only makes instructional variations possible but also allows the teacher to create ten different formats for activity sheets or ditto masters. Of these ten, six allow the child to work independently and four require the presence of the teacher. These will be described in the section on material modification.

Exhibits 9–2 and 9–3 contain illustrations of two different instructional groupings. These may vary in the degree of preparation or material identification and require different amounts of teacher time. Exhibit 9–2 shows a class of 23 children divided into three groups. Each group is involved in a different topic. The class period is 45 minutes long and the teacher wishes to spend about 15 minutes with each group. This requires that the teacher plan the activities so that his or her presence is provided to the proper group at the appropriate time. It also means that the teacher has to plan for a certain amount of independent work for the group he or she is not with.

The geometry group will conduct its teacher-directed activities manipulatively and will work with two different activity sheet formats, display/identify and write/identify. The fraction group will conduct its direct-instruction activities through manipulative inputs and spoken outputs and will work independently in display/write and write/write interactions. The addition group will conduct its directed-instruction activity aurally and will work independently in write/write and display/write formats.

The grouping illustrated in Exhibit 9–3 varies only in organization. The same topic will be covered in the same manner with the same materials. The only instructional variation will be that the teacher will move from group to group and switch assignments with each move. Each group will have one direct instructional period with the teacher and two independent activities.

The directed activity will be conducted in aural language (e.g., the teacher will say, "What is two times three?" and the learner will respond, "six"). The written activities will consist of multiplication items (e.g., $2 \times 3 = 6$) and the learner will write the answer. The display/write activities will resemble the following: Draw a path to connect the correct array with the written answer.

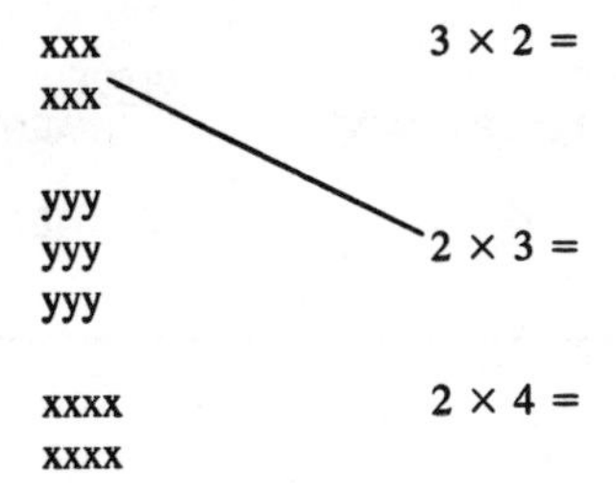

One way to facilitate the use of the IU is to have both regular teachers and special education teachers work together to conduct instructional activities and to

Exhibit 9–2 Instructional Grouping I

	GROUP A - GEOMETRY 8 STUDENTS	GROUP B: FRACTIONS 10 STUDENTS	GROUP C: ADDITION 5 STUDENTS
15 MINUTES	MANIPULATE/MANIPULATE * INPUT: TEACHER WALKS THE PERIMETER OF A GEOMETRIC SHAPE OUTPUT: LEARNER DOES THE SAME	DISPLAY/WRITE INPUT: WRITE THE FRACTION THAT NAMES THE SHADED PART. OUTPUT: LEARNER WRITES $\frac{1}{2}$	WRITE/WRITE INPUT: $3 + 2$ WRITE THE ANSWER, OUTPUT: LEARNER WRITES 5
15 MINUTES	DISPLAY/IDENTIFY INPUT: FROM THE CHOICES, MARK THE SHAPE THAT IS THE SAME AS THE FIRST SHAPE OUTPUT: LEARNER MARKS:	MANIPULATE/SAY * INPUT: TEACHER REMOVES PORTION OF SHAPE AND ASKS LEARNER TO NAME THE PART OUTPUT: LEARNER SAYS: "ONE FOURTH"	DISPLAY/WRITE INPUT: WRITE THE NUMBER THERE IS IN ALL OUTPUT: LEARNER WRITES: 5
15 MINUTES	WRITE/IDENTIFY INPUT: CIRCLE MARK THE SHAPE THAT SHOWS THE WORD. OUTPUT: LEARNER MARKS: CIRCLE	WRITE/WRITE INPUT: ONE HALF WRITE THIS WORD STATEMENT AS A NUMERAL. OUTPUT: LEARNER WRITES: $\frac{1}{2}$	SAY/SAY * INPUT: TEACHER SAYS: "I AM GOING TO SAY SOME ADDITION ITEMS. SIX PLUS SIX. TELL ME THE ANSWER." OUTPUT: LEARNER SAYS: "TWELVE"

* TEACHER PRESENT IN GROUP

Exhibit 9–3 Instructional Grouping II

CONCEPT: MULTIPLICATION

	GROUP A - 8 STUDENS	GROUP B - 10 STUDENTS	GROUP C - 5 STUDENTS
15 MINUTES	SAY/SAY * INPUT: TEACHER SAYS, "WHAT IS TWO TIMES THREE?" OUTPUT: LEARNER SAYS: "SIX"	WRITE/WRITE INPUT: 2X3 = WRITE THE ANSWER. OUTPUT: LEARNER WRITES: 2x3 = 6	DISPLAY/WRITE INPUT: DRAW A PATH FROM THE ARRAY TO THE WRITTEN STATEMENT. 3X2 2X3 2X4 OUTPUT: LEARNER DRAWS PATH. 3X2 2X3 2X4
15 MINUTES	DISPLAY/WRITE INPUT: DRAW A PATH FROM THE ARRAY TO THE WRITTEN STATEMENT. 3X2 2X3 2X4 OUTPUT: LEARNER DRAWS PATH. 3X2 2X3 2X4	SAY/SAY * INPUT: TEACHER SAYS, "WHAT IS TWO TIMES THREE?" OUTPUT: LEARNER SAYS: "SIX"	WRITE/WRITE INPUT: 2X3 = WRITE THE ANSWER. OUTPUT: LEARNER WRITES: 2x3 = 6
15 MINUTES	WRITE/WRITE INPUT: 2X3 = WRITE THE ANSWER. OUTPUT: LEARNER WRITES: 2x3 = 6	DISPLAY/WRITE INPUT: DRAW A PATH FROM THE ARRAY TO THE WRITTEN STATEMENT. 3X2 2X3 2X4 OUTPUT: LEARNER DRAWS PATH. 3X2 2X3 2X4	SAY/SAY * INPUT: TEACHER SAYS, "WHAT IS TWO TIMES THREE?" OUTPUT: LEARNER SAYS: "SIX"

* TEACHER PRESENT IN GROUP

prepare or adapt instructional materials. Assume that a child is enrolled in a regular class. The topic being presented is single-digit multiplication concepts. The regular teacher has the workbook or textbook, which is primarily composed of written symbolic items. One page has a number line. The special education teacher could work manipulatively with the learner to develop the concepts of multiplication as equal sets, expressions in the form of arrays, and points on a number line. The line could be made with dots placed on the floor and the youngster could be given demonstrations showing the teacher walking or moving objects a certain number of dots at a time.

Material Modification

Instructional materials are modified for use with learning-disabled children to:

1. partial out the efforts of one disability (e.g., reading) on another (e.g., word problems)
2. extrapolate the presentations of a skill or concept over a longer period of time than is possible with the original materials
3. represent concepts or skills differently
4. increase or decrease the number of repetitions for a given set of items
5. provide teachers with varying means of interacting with individual learners

Exhibit 9–4 shows a page of fraction items. Exhibits 9–5 and 9–6 illustrate two modifications of the ten possible worksheet formats that could be undertaken for Exhibit 9–4.

Each of these adaptations could be prepared in multiples, the number undertaken being a function of the needs of the individuals. One might use square regions at one time and circular regions another. The number of items could be increased or decreased according to the needs of the learners.

As professionals, we generally possess the knowledge and capability to modify instructional practices and materials. What we lack is the availability of resources (e.g., time) to effect these adaptations.

DEVELOPMENT

Development is defined as the process of creating and producing. While the process of development is not intellectually complex, it is time consuming, demanding, and arduous. It is unlikely that many local education agencies can effectively develop curriculum and instructional materials that are sufficiently comprehensive to meet the needs of learning-disabled children. The cost of planning, writing, producing, duplicating, evaluating, and disseminating is gener-

Exhibit 9–4 Worksheet of Fraction Items (Write/Manipulate Interaction)

INSTRUCTIONS: The name of a proper fraction is written in each frame. Use scissors paper, and paste to make a picture representing that fraction. (Extra points will be awarded for creativity, but be sure to show the fraction.)

ACTIVITY NUMBER

1. $\frac{2}{4}$	2. $\frac{2}{3}$
3. $\frac{3}{4}$	4. $\frac{2}{5}$

Exhibit 9–5 Worksheet of Fraction Items Showing Display/Identify Interaction

INSTRUCTIONS: In each row there is a picture and some choices. Mark the choice that matches the first picture. The first one is done as an example.

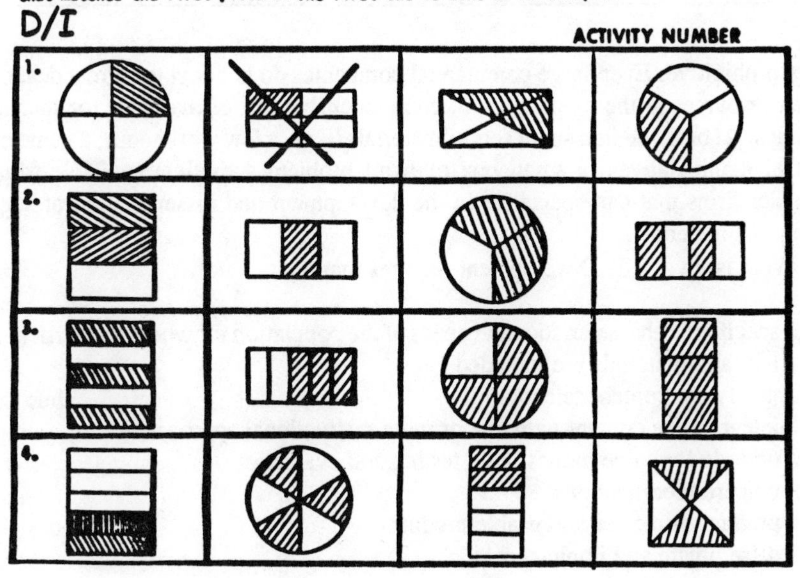

Exhibit 9–6 Worksheet of Fraction Items Showing Display/Manipulate Interaction

INSTRUCTIONS In each frame there is a picture which shows the meaning of a proper fraction. In the rest of the frame, use the materials your teacher gives you to show that same fraction.

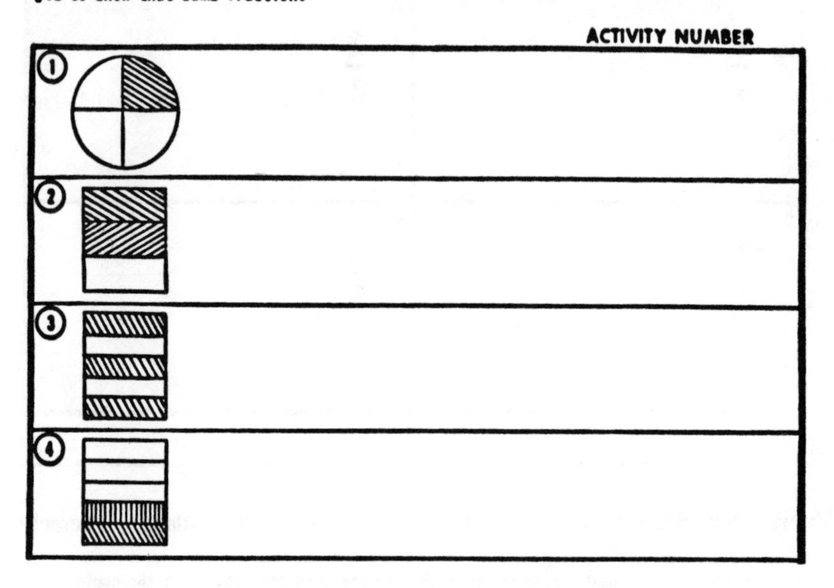

ally prohibitive. Even large commercial companies do not devote a great deal of their resources to the preparation of comprehensive sets of materials for special groups. At best, we find small sets of materials (e.g., a few workbooks, a game or set of manipulatives, a small box of word problem activities) available from smaller firms that can specialize in the development and dissemination of segments of materials.

What is involved? Development requires that one:

1. specify the characteristics and needs of the population for whom the curriculum and materials are intended
2. justify the content selection
3. integrate the content with one or more instructional approaches
4. provide for some modest field testing and evaluation
5. prepare modifications
6. produce attractive and usable products
7. disseminate and implement

Experience with *Project MATH* (Cawley, Goodstein, Fitzmaurice, Lepore, Sedlak, & Althaus, 1977) and *Multi-Modal Mathematics* (Cawley, Fitzmaurice, Shaw, Schummann, & Norlander, 1980) shows that any comprehensive endeavor takes four to six years to develop and produce. It requires a core group that can absorb staff turnover and more important, resist the temptation to continuously change with every new idea. Some balance and perspective is needed or the project will never be finished. Unfortunately, it is only the people who are intimately involved who have a true picture of the changes needed. Many products are developed that contain suggestions about what to do and how to do it that have long ago been proven inadequate.

With respect to mathematics, perhaps the most ill-advised of these is the recommendation to use cue words (e.g., *left, take away, divide*) as an aid in the selection of the operations for a word problem. Even more baffling is the use of color codes that heighten attention to cue words. Cue words have long-term negative effects. Examine the following:

A. A boy has 9 apples.
He gave 3 away.
How many does he have left?
B. A boy has 9 apples.
He gave 3 away.
How many does he have now?
C. A boy has 9 apples left after
he gave away 3. How many did
he start with?

Example A stresses the cue word.
Example B eliminates the cue word in the question.
Example C illustrates the confusion created when one depends on cue words.

As can be seen, cue words encourage rote computational habits, not problem solving. The proper adaptation is to eliminate their use.

IMPLEMENTATION

Unless there is effective and meaningful implementation, the processes of selection, adaptation, and development are wasted. A district needs to have an effective plan to implement the processes we have discussed. The education of learning-disabled children cannot be left to the individual interests of the various teachers, diagnosticians, and researchers who serve them and work with them.

Clearly, implementation requires study, planning, presentation, and promotion. The planning and budget needs of the school districts must also be met. Ancillary services can be provided when necessary, along with guidance in fulfilling the responsibilities of educators as outlined in Public Law 94-142.

REFERENCES

Ballew, H., & Cunningham, J. (1982). Diagnosing strengths and weaknesses of sixth-grade students in solving word problems. *Journal for Research in Mathematics Education, 13,* 202–210.

Cawley, J.F., Fitzmaurice, A.M., Shaw, R.A., Schunmann, J., & Norlander, K. (1980). *Multi-model mathematics.* Unpublished manuscript, University of Connecticut, Storrs.

Cawley, J.F., Goodstein, H.A., Fitzmaurice, A.M., Lepore, A., Sedlak, R., & Althaus, V. (1976). *Project MATH.* Tulsa, Okla.: Educational Progress Corporation.

Fitzmaurice, A.M. (1980). LD teacher's self-ratings on mathematics education competencies. *Learning Disability Quarterly, 3,* 90–94.

Flournoy, R. (1960). Meeting individual differences in arithmetic. *The Arithmetic Teacher, 7,* 80–86.

Gall, M.D. (1981). *Handbook for evaluating and selecting curriculum materials.* Boston: Allyn & Bacon.

Lambie, R.A. (1980). A systematic approach for changing materials, instructions and assignments to meet individual needs. *Focus on Exceptional Children, 13,* 1–12.

Vale, C.A. (1980). *National needs assessment of educational media and materials in handicapped.* Princeton, N.J.: Educational Testing Service.

Chapter 10

Classroom Management Tactics: Pre-K through Secondary

Elizabeth McEntire

It is not uncommon to find the term *classroom management* equated with the teacher's disciplinary role. However, a more comprehensive definition broadens this role to that of

> . . . conductor of the orchestration of classroom life: planning curriculum, organizing procedures and resources, arranging the environment to maximize efficiency, monitoring student progress, anticipating potential problems. (Lemlech, 1979. p. 5)

Orchestration of the mathematics classroom, like any other classroom, requires a great deal of teacher knowledge about the content to be learned, the unique characteristics of each learner, interpersonal communication, the effective use of group power, and the best ways to facilitate self-direction.

Most mathematics teachers are adequately trained in the content of the discipline but may lack sufficient information concerning learner differences that affect learning behaviors. These individual differences are the general variables that play an important part in the preplanning necessary for good classroom management.

GENERAL VARIABLES IN CLASSROOM MANAGEMENT

Learner differences in levels of cognitive development, cognitive style, learning style, or deviance from classroom norms for learning and behavior are easily recognized as major factors in managing classroom groups. However, teachers also vary. In approach they may favor teacher-centered over learner-centered management, or competitive over cooperative activities. All of these variables contribute significantly to the "goodness of fit" (Laten & Katz, 1975) between the learner and the learning environment.

Levels of Cognitive Development

The influence of developmental psychology on the mathematics classroom can be seen in current research (Case, 1978; Hiebert, Carpenter, & Moser, 1982; Souviney, 1980) relating mathematical activities to levels of cognitive development. The results of such studies have led to a general recognition of "readiness" as a condition that should be present prior to introducing students to the abstractness of mathematical concepts and language. Delaying the introduction of certain mathematical topics until cognitive competency has developed is suggested as a means of improving the efficiency of instruction and lessening the anxiety and phobia often associated with mathematics learning (Hamrick, 1980; Souviney, 1980).

It is now generally accepted that learners "know" mathematics at different cognitive levels and that these follow an invariant developmental sequence: (1) the inactive, instrumental, intuitive, sensory-motor, nonverbal level; (2) the iconic, representational, perceptual, empirically abstractive, mental imagery level; and (3) the symbolic, reflectively abstracting, mental coding manipulation level (Bruner, 1963; Gallagher & Reid, 1981; Sharma, 1979).

The progression of individual students through these levels of cognitive development has two implications for classroom management. First, the ways of presenting mathematical concepts, particularly the student activities, must be selected in relation to the developmental level of the learner. The teacher who provides a selection of activities including active manipulation of concrete objects and pictorial or model representations of the problem in addition to the usual symbolic tasks of the mathematics textbook or worksheets allows for individual differences in cognitive development as each new concept is taught. When such variations in classroom tasks are available to all learners, students can successfully move from activities at the beginning developmental level to those at the higher level without the fear of failure and lack of adequate opportunity to develop understanding.

A second implication for classroom management related to levels of cognitive development is the use of the developmental level as a criterion for grouping learners. Developmental placement allows students to be grouped differently for different mathematical topics and provides flexibility in moving students to higher-level tasks as soon as readiness is indicated (Sharma, 1979).

Cognitive Styles and Learning Styles

Psychologists and educators interested in individual differences among students in a given classroom have studied a number of personality and cognitive factors. One such factor, called "cognitive style," appears to bridge such personality-cognitive dimensions of students (Sigel & Coop, 1974). The term is used to describe differences among children or adults in their perception and categorization of the environment and in their response in particular situations. Individual differences in cognitive style may occur in both the approach to acquiring new information and in personal selections of problem-solving strategies and responses. Although these differences are not attributable to intelligence, they do appear to have significant bearing on the individual learner's approach and response to specific types of mathematical tasks, teaching methods, or classroom structure.

Variations in cognitive style have been described along several dimensions:

- *Convergent vs. divergent thinkers.* Convergent thinkers tend to reach conclusions based upon learned conventional forms of reasoning, which lead to a commonly accepted and expected response. Divergent thinkers, in contrast, appear to produce multiple, logically possible solutions through unique, creative thinking (Rice, 1979).
- *Scanning vs. focusing attenders.* Students who scan tend to uniformly attend to all the details of a situation, including the incidental ones. Students who focus, by contrast, fix their attention on a few key details, ignoring peripheral cues, even when important (Rice, 1979).
- *Descriptive/analytic vs. relational/contextual vs. categorical/inferential categorizers.* Descriptive/analytic categorizers build classes on objective, observable characteristics that are part of the total situation. Relational/contextual categorizers group items together based on functional, temporal, or spatial contiguity. Categorical/inferential categorizers form categories based on inferences made about the objects grouped together (Kagan, Moss, & Sigel, 1963).
- *Impulsive vs. reflective conceptual tempo.* Impulsive tempo is used to describe rapid responders who make many errors. Reflective tempo describes slower responders who tend to make fewer errors. Differences in conceptual tempo may be related to cognitive control in delaying responses rather than personal preference (Kagan, Rosman, Day, Albert, & Phillips, 1964).
- *Field-independent vs. field-dependent attenders.* Field-independent attenders are able to overcome the effects of distracting background elements in attending to the relevant aspects of a learning situation. They are more differentiating and analytical, more socially impersonal, more internal in locus of control, and need few, if any, external cues to organization of information. Field-dependent attenders, in contrast, are more global and undifferentiating in their attention, more socially sensitive, more external in locus of control, and more dependent on external organization and structuring of information (Witkin, Moore, Goodenough, & Cox, 1977).

Differences in cognitive style may be both developmental and preferential. Such differences have been suggested as an explanation for why children of comparable IQ scores differ so widely in interests in classroom activities and in performance levels on various mathematical tasks. In managing classroom behavior the teacher who is knowledgeable of cognitive style differences can vary tasks, materials, or teaching style as needed. Also, cognitive style differences, such as impulsive tempo, may be modified by planned interventions (Sigel & Coop, 1974).

In addition to cognitive style differences, Dunn and Dunn (1972) suggest that individual differences in "learning styles" should be diagnosed for effective instructional planning. Teacher observations of the 12 variables listed below provide a foundation for individualization or small homogeneous group arrangements that facilitate "a good match" between the mathematical learning environment and the learner.

1. Time	When is the student most alert? In the early morning, at lunchtime, in the afternoon, in the evening, at night?
2. Schedule	What is the student's attention span? Continuous, irregular, short bursts of concentrated effort, forgetting periods, etc.?
3. Amount of Sound	What level of noise can the student tolerate? Absolute quiet, a murmur, distant sound, high level of conversation?
4. Type of Sound	What type of sound produces a positive reaction? Music, conversation, laughter, working groups?
5. Type of Work	How does the student work best? Alone, with one person, with a small task group, in a large team, a combination?
6. Amount of Pressure	What kind of pressure (if any) does the student need? Relaxed, slight, moderate, extreme?
7. Type of Pressure and Motivation	What helps to motivate this student? Self, teacher expectation, deadline, rewards, recognition of achievement, internalized interest, etc.?
8. Place	Where does the student work best? Home, school, learning centers, library media corner?

9.	Physical Environment and Conditions	Floor, carpet, reclining, sitting, desk, temperature, table lighting, type of clothing, food?
10.	Type of Assignments	On which type of assignments does the student thrive? Contracts, totally self-directed projects, teacher-selected tasks, etc.?
11.	Perceptual Strengths and Styles	How does the student learn most easily? Visual materials, sound recording, printed media, tactile experiences, kinesthetic activities, multimedia packages, combinations of these?
12.	Type of Structure and Evaluation	What type of structure suits this student most of the time? Strict, flexible, self-determined, jointly arranged, periodic, self-starting, continuous, occasional, time-line expectations, terminal assessment, etc.? (Dunn & Dunn, 1972, pp. 29–30)

Behavior

Within any classroom there are certain learners who are significantly different from the teacher's expectations. Their behavior might include task refusal, excessive inattentiveness, aggressiveness, or social withdrawal. The teacher's perception of the severity of these learning and behavioral differences is frequently related to the teacher's feelings of competence in coping with them.

However, certain observations made by teachers seem to be indicators of severe learning and behavioral problems that require special management and/or additional teaching assistance. Severe learning problems are associated with an extensive need for one-to-one, teacher-student instruction and alternatives to the regular grade-level curriculum, even when activities are available for the various levels of cognitive development (DeLoach, Earl, Brown, Poplin, & Warner, 1981). Severe behavioral problems are indicated by one or more of the following characteristics exhibited over a long period of time and to a marked degree that adversely affect educational performance:

- an inability to learn that cannot be explained by intellectual, sensory, or health factors
- an inability to build or maintain satisfactory interpersonal relationships with peers and teachers
- inappropriate types of behavior or feelings under normal circumstances

- a general pervasive mood of unhappiness or depression
- a tendency to develop physical symptoms or fears associated with personal or school problems (Bower, 1969)

Students who exhibit these indicators of severe learning and behavioral problems may become a source of classroom disruption if management planning does not address their particular needs.

Teacher-Centered vs. Learner-Centered Management

Although teachers frequently state their desire to see learners develop more self-direction, classroom management strategies often contradict such statements. Classroom instruction dominated by teacher talk, teacher choice, and teacher reinforcement does little to teach students the lifelong value of self-directed learning. Teacher-centered classrooms often ignore student needs to be active learners, to make choices, to be social, and to feel accepted, confident, and secure (Wallen & Wallen, 1978). The teacher who deliberately creates a learner-centered, communicating atmosphere manages a classroom that promotes student growth in the following:

- assuming responsibility
- feeling accepted and respected
- self-motivation
- active participation
- human interaction
- feeling secure enough to respond and inquire
- feeling understood
- becoming self-disciplined
- verbalizing with ease
- achieving insights
- becoming aware of appropriate attitudes
- changing values
- responding to genuineness in others
- respecting and valuing interaction with the teacher (adapted from Pine & Boy, 1977, pp. 14–17)

Learner-centered classroom management focuses the teacher's attention on (1) evidencing respect for the learner; (2) providing opportunities for student choices and accepting them; (3) concentrating on the needs, problems, and feelings of students; (4) emphasizing good communication skills; and (5) allowing students the time and freedom to explore and learn self-direction (Pine & Boy, 1977). Within this framework of interpersonal interactions the teacher provides mathematical content and activities to enhance growth of self-management.

Competitive vs. Cooperative Activities

Given the importance of cooperative behavior for adult success, it might be assumed that schools would emphasize all types of cooperation in the learning environment. The fact is, however, that the current emphasis in most schools and classrooms is on constant comparisons among students and the use of competition as a motivational device. Classroom management that emphasizes competition, whether for teacher attention, tokens, grades, or other rewards ensures that:

1. students will provide little help to one another and be inept in engaging in helping behavior;
2. students will form a pecking order, based largely on academic ability;
3. students with innate academic ability may often be reluctant to exercise that potential in the face of the jealousy of their peers;
4. students who are always at the bottom of the class will seek to avoid, disrupt, or simply tune out the classroom experience; and
5. teachers will be uninterested in socializing students to practice helping behaviors toward each other. (Slavin, 1981, p. 1)

Cooperative learning methods have been developed by a number of researchers. The Jigsaw Method (Aronson, 1978) has each student in a six-member group learn a unique piece of information, discuss it with counterparts in other groups, and then teach it to his or her team. Two other methods, Teams-Games-Tournament (TGT) and Student Teams Achievement Divisions (STAD) (Slavin, 1980), emphasize group competition. In these methods students first hear a teacher presentation, then study materials in four- or five-member heterogeneous teams, and then take a quiz (STAD) or play an academic game (TGT) to show what they have learned. Team scores are recorded and rewarded.

Research results (Slavin, 1981) indicate that academic performance following cooperative learning methods was at least as high if not higher than in competitive learning. Projected long-term consequences of cooperative methods include (1) students becoming more peer oriented, more altruistic; (2) students becoming more accepting of classmates different from themselves; (3) students broadening their friendship networks, thus decreasing the isolation and alienation of certain subgroups; and (4) students developing good mental health, i.e., increasing self-esteem.

Traditional classroom management emphasizes competitive, individualistic learning based on grading systems, tests, quizzes, objective exams, or statistical norms. Cooperative learning methods encourage group members to help each other on learning tasks to achieve both group and individual goals.

MANAGING INSTRUCTION

Although instructional approaches to mathematics are always peculiar to each particular teacher, two general approaches having significant management implications can be identified: direct instruction and mathematics laboratories. The structured, direct-instruction classroom is managed quite differently from the mathematics laboratory.

The Structured Direct-Instruction Classroom

The structured direct-instruction classroom has been clearly described by Silbert, Carnine, and Stein (1981). Direct instruction may vary depending upon the student being taught and the specified objectives. For example, average students in the intermediate grades would have a great deal of individual, independent seatwork while students who have encountered difficulty would have more structured, more teacher-directed instruction.

The first aspect of organizing instruction is represented by the following eight-step instructional design:

1. State the teaching objectives specifying what the students will be able to perform.
2. Devise the problem-solving strategies to be taught.
3. Determine any necessary preskills for the performance objectives.
4. Sequence the performance skills in an optimum order to facilitate their acquisition.
5. Select a teaching procedure appropriate to the task and the learners, e.g., modeling, leading the student, or questioning to solicit and train correct responses.
6. Design a specific format for the lesson presentation including (1) what the teacher says and does; (b) examples to be presented with expected responses; and (c) correction procedures for errors.
7. Select problem examples of currently or previously taught objectives for group work or individual seatwork.
8. Provide practice and review, beginning with massed practice of new objectives and moving to systematic review.

The second aspect of direct instruction is the teacher's selection of presentation techniques. Techniques are chosen to maintain student attention and participation during group instruction and to ensure that students master all skills presented. Suggestions for managing attentiveness include the following:

- Monitor the length of teacher presentation or explanation by continuously observing students.
- Provide frequent opportunities for student response during teacher presentation (responses may be individual or group unison).
- Use pretaught signals as cues for response time.
- Pace instruction and responses by a lively animated, nonhesitant manner.
- For small group instruction, seat students in a semicircle facing the teacher, backs to the rest of the class.
- Place distractible students toward the middle of the semicircle, in front of the teacher, but not together.

Direct instruction is a procedure for teaching to a specific performance criterion. Such criterion teaching requires that the teacher be actively involved in constantly monitoring student performance, in order to diagnose and remediate errors through correction procedures (i.e., increasing attention through reinforcement, analyzing computational errors, interviewing students to determine faulty thinking, or reteaching skills).

The third and final aspect of direct instruction involves testing and placing students in instructional groups to ensure effective use of resources and time. Placement tests provide the basis for initial homogeneous grouping beginning with the formation of the low performance group and moving upward. Moving a group through the curriculum of objectives is the critical teaching skill. Ideally, before moving on, all students in the group should be able to respond correctly to 85–90 percent of the problems selected for drill and review with no teacher assistance. A two-day interval is suggested for the introduction of new skills. Students who consistently require more time may be moved to a lower group or individually tutored.

The Mathematics Laboratory

The mathematics laboratory approach refers to both a particular type of room designed for carrying out individual or group experiments and particular types of mathematical activities designed to provide the learner with direct experience in "doing mathematics." Such experiences may include performing an experiment, viewing a film, playing a game, discussing a mathematical problem, reading about mathematics, programming a computer, building a model, constructing a problem solution, making a survey, drawing designs, making graphs of information, role playing a mathematical problem situation, or just quiet thinking (Johnson & Rising, 1972).

Work in the mathematics lab is centered on problem solving, discovery learning, and constructive thinking. Lab activities frequently are designed to teach students the applications of mathematical concepts and skills in real-world situations. The wide variety of activities available provide success opportunities for students differing in cognitive development or other individual variables.

Obviously, the management of the mathematics lab involves more spontaneous decision making on the part of the teacher than in the structured direct-instruction classroom. Students are given a great deal of responsibility in selecting materials, learning to utilize equipment, moving freely within the workspace, discussing activities with other students, and keeping the classroom orderly and businesslike (Kidd, Myers, & Cilley, 1970).

The teacher's initial responsibility is to procure all the types of materials needed for the wide range of activities available in the lab. Materials and equipment must be organized and stored in such a way that students can easily find, use, and return them. A second major task for the teacher is the preparation or procurement of guidesheets for the various activities that serve as the instructional guides. Teaching consists of demonstrating experiments while questioning students about their observations; scheduling independent use of equipment; supervising and guiding students as they work; listening to students explain their procedures and findings; and evaluating each student's attitude, work habits, and accomplishments. Student progress may be demonstrated by a specific project product or through oral communication between student and teacher.

Frequently, teachers mix the two approaches we have described by utilizing direct instruction in the laboratory setting with students who benefit from structured learning and cannot demonstrate the self-management necessary for the laboratory approach. It can also prove beneficial to students to present some types of mathematical skill acquisition through direct instruction with practice while utilizing laboratory experiments for basic mathematical concepts and "real-world" mathematics applications.

MANAGING THE INSTRUCTIONAL GROUP

A most important component of classroom management is the psychological dynamics of the classroom group (Wallen & Wallen, 1978). The teacher's knowledge of these dynamics will affect successful management of individual disruptive students as well as the teacher's ability to work *with* the classroom group

Groups have a personality of their own which can be "healthy" or "unhealthy." The effective classroom manager understands the determinants ot the group atmosphere and is competent in selecting teaching strategies to change group dynamics when necessary. Likewise, the effective manager recognizes that classroom groups are unique in that (1) the group is formulated for the goal or purpose of learning; (2) group goals and group participation are mandatory; and (3) the members of the group have no control over the selection of the leader (assuming this is the teacher's role) and little recourse from poor leadership (Getzels & Thelen, 1960).

The conditions that determine the health of the group are directly affected by the teacher's personal influence and, indirectly, by the teacher's interactions with group cohesiveness, standards, and structure.

Teacher's Personal Influence

Several conditions appear to determine the teacher's personal influence on students: (1) influence is positive and effective if the teacher is liked; (2) influence is more powerful if students respect the teacher's knowledge, teaching competency, and status as an adult; (3) personal influence decreases when used excessively; and (4) influence is increased when individualized instruction allows all students to maximize rewards and minimize punishments.

When increased personal influence is needed by the teacher, certain changes in behavior may be helpful. Emotional ties between students and teachers may be strengthened by teacher friendliness, support, and kindness combined with consistency and firmness. In addition, personal influence is strengthened when teachers find new ways to demonstrate their knowledge of the content and their competence in teaching it. Giving students an opportunity to observe nonclassroom expertise (e.g., competency in athletics, playing an instrument, etc.) may also increase student respect. Incidentally, student confidence in the teacher is also related to the teacher's ability to admit a lack of knowledge or the fact that he or she has made a mistake. Finally, the teacher's use of methods other than personal influence to get students to do things increases the potential power of that influence. Introducing intrinsically motivating assignments, incentive techniques, or group influence to manage individual problems preserves the power of the teacher's personal influence.

Group Cohesiveness, Standards, and Structure

In addition to the teacher's personal influence, classroom management is directly affected by the group influence that develops. Three characteristics of all groups seem to make the difference between "healthy" and "unhealthy" group influence: (1) the group's cohesiveness or feeling of "oneness," (2) the group's selection of standards or rules, and (3) the group's structure of statuses and subgroups that influence individual members. Healthy classroom groups may be described in the following fashion:

- Cohesive groups work more efficiently because little time is spent in discord.
- Cohesive groups put pressure on individual members to conform to group standards.
- Group pressure on deviant members is indicated by the type and number of communications addressed to them.
- Cohesiveness can be increased by stressing the ways in which the group meets an individual's needs for being an active learner, socializing, or feeling confident and secure.
- Cohesiveness may be increased by building the prestige of low-status members (e.g., give them tasks or rewards desired by other group members).
- Cohesiveness may be increased by designing cooperative activities for the group and increasing interaction among group members.
- Cohesiveness may be increased by distinguishing this group from other groups (e.g., comparing this classroom group with another) (Wallen & Wallen, 1978).

Students who are not attracted to classroom group membership and are not influenced by the group detract from cohesiveness. Frequently such students fall into one of the following categories: (1) rejected by the group (i.e., they have concluded they will never be liked and accepted); (2) egocentric (i.e., they are unable to contribute to others and seek only to meet their own needs): (3) young "old-timers" (i.e., they find adult relationships more attractive than peer relationships); and (4) anomic individuals (i.e., they are disoriented from normal classroom groups but often have group affiliations outside the classroom). These individuals may require the use of the teacher's personal influence to initiate motivation to become part of the group.

Group standards or rules are the behaviors that a cohesive group pressures its members to exhibit. If group standards contradict teacher standards for behavior, the teacher may work toward reducing group cohesiveness by using the opposite of procedures previously mentioned for building cohesiveness. Once cohesiveness is reduced, standards may be changed by teacher-planned classroom discussions that allow the free expression of opinion and argumentation about classroom standards.

Group structure refers to the system of social stratification whereby individual members have a certain status in the group and are thus involved in certain patterns of interaction. Status, based on such criteria as sex, age, personality, size, health, looks, material possessions, athletic ability, academic ability, self-discipline, ethnicity, or social class, determines an individual's influence with other group members. Status also affects peer and teacher judgments of academic or social behavior (for example, behaviors are acceptable when exhibited by high-status group members but not acceptable when exhibited by low-status members).

Location in the group structure plays a major role in the individual's degree of autonomy and vulnerability to control by others. The higher the student's status in the group the more willing he or she is to be influenced by the group. Low-status members deviate from group standards more often and are less easily influenced, which leads to rejection and decreased communication. The formation of subgroups or cliques of high-status or low-status group members produces decreased cohesiveness and increases the teacher's management problems.

Group structures can be altered, however, by providing opportunities for social mobility, such as raising a person's status, or by having cooperative group activities where members work together toward a common goal. Social mobility of low-status students can be encouraged by having other students see them doing good things or by increasing their opportunities for interaction with group members.

A major need of any student in the classroom is peer group acceptance. The mainstreaming of many mildly handicapped students has brought them into new classroom groups. If such students have been ostracized, branded, or stereotyped because of their special education label (i.e., Learning Disabled, Emotionally Disturbed, Mentally Retarded) they have even greater need for status, respect, and security in the classroom group. Teachers often prepare regular classroom groups for special students by focusing discussions on ways all individuals differ and ways all individuals are alike. Classroom groups can be given a chance to ask questions and build understanding of how it feels to be labeled as handicapped. The purpose of these activities is always directed toward the acceptance of individual differences and the provision of high group status, not to foster pity for the incoming student (Lemlech, 1979).

Learner-centered classroom management requires that the teacher plan a consistent movement from use of direct personal influence to use of group influence. Although younger children may need the modeling of teacher direction and influence, the goal of management strategies should change as the student progresses in school, becoming more focused on peer group influence and self-direction. Specific types of management techniques therefore may be most effective with various school-age groups.

MANAGING THE PRESCHOOL MATHEMATICS CLASSROOM

Mathematics programs currently utilized for young children are quite diverse in philosophy and content (Johnson & Wilson, 1976). The three approaches to mathematics commonly found in preschool programs require very diverse management strategies. What is to be taught and how it will be taught differ among behavioral, cognitive-developmental, and naturalistic approaches to mathematics programs for young children.

Behavioral Programs

The behavioral approach is a form of direct instruction. Teaching is primarily didactic and the focus of instruction is a carefully planned sequence of tasks to teach mathematical facts, concepts, and principles. Using an expository method, the teacher directs the students in both oral and written drill of correct responses to mathematical questions or problems. Lesson plans are highly structured, including both the teacher's instructions and the children's expected responses. Teaching materials often suggest teacher rewards for appropriate individual behavior or for full participation of all students during each presentation. Both teacher presentation and drill procedures are rapidly paced to hold the short attention of the young child.

Cognitive-Developmental Programs

Another approach to preschool mathematics is represented by the discovery/problem-solving activities of the cognitive-developmental curriculum. Its focus in mathematics instruction is upon experiments that allow the young learner to actively experience the mathematical reasoning of classification, seriation, spatial relationships, etc. This approach is best carried out in a lab setting, though it requires extra time to teach self-directed behaviors for managing equipment, for appropriate social interactions, and for self-report of learning.

Naturalistic Programs

The naturalistic approach to mathematics programming is, perhaps, unique to the preschool level. The naturalists advocate a child-centered curriculum with the teacher as a guide or facilitator. Curriculum is developed from the naturally occurring interests of the child. Mathematics instruction or exploration occurs as a function of the need for it in connection with an activity unit or a child's natural curiosity. While this approach is somewhat unplanned and unstructured, the preschool teacher who is knowledgeable about mathematics finds many spontaneously occurring events that present mathematical information or problems.

Classroom management in this approach is a highly flexible interaction between teacher questioning and child responding in both the spontaneous planning of mathematical activities and in carrying them out. Child-selected mathematics activities are structured by the teacher to meet children's needs for physical activity, social interactions, play, construction, and just thinking.

General Management Techniques

Regardless of the curricular approach selected, certain general statements are applicable to the management of the preschool mathematics classroom:

- Teachers who genuinely like children and respect them as individuals *look like* they do. They smile easily and demonstrate interest, enthusiasm, and physical affection.
- Teacher credibility (i.e., saying what they mean and meaning what they say) is an essential part of children's needs for structure and dependability.
- When situations require rules, children should participate in the wording of the rules and in developing a rationale for the rules so that a child can easily memorize and understand them.
- Teachers should not do things for students that children can do for themselves with some teacher instruction and demonstration.
- Planning, scheduling, and pacing activities should minimize inactive wait time for young children.
- Learning is easier and more pleasant for the child when teachers give positive cues to appropriate behavior (i.e., tell the child what to do rather than what not to do) and when teachers praise appropriate behavior.
- Teacher praise in simple, declarative statements that name the specific behavior and use a wide variety of words and phrases is most effective in helping young children discriminate appropriate behavior (e.g., "I'm really proud of the way you figured out that problem," etc.).
- Praise accompanied by nonverbal communications of approval such as a smile or pat is most rewarding.
- Children can be helped in focusing and holding attention if the teacher:

    1. does not begin before the whole group is attending
    2. states an expectation that each child will attend fully at all times
    3. keeps the lesson moving at a good pace
    4. constantly scans the group to monitor attention
    5. uses a variety of questions and response demands and avoids repetition
    6. questions before naming a child to answer and ensures that all children participate whether they volunteer or not
    7. changes voice tone, emphasizes key words, or provides transitional signals such as "all right" to keep attention focused

8. models paying attention when children are speaking
9. models the kinds of observations and thought processes the children should be using when attending by thinking aloud, making comments about what may happen next, or describing and commenting on children's responses
10. monitors the attention span of young children well enough to end an activity too early rather than too late (Brophy, Good, & Nedler, 1975)

Management of the preschool mathematics group requires a great deal of teacher oral direction and use of personal influence as the young child learns to differentiate school-appropriate behavior from behavior appropriate for other environments. Two management techniques are particularly applicable to preschool classroom groups.

Internalized Speech

In his studies of the development of socialized behavior in children, Vygotsky (1977) noted the young child's overt self-talk in making decisions about behavior. Young children often repeat aloud directions or warnings previously heard from adults when they face a behavioral decision. This overt self-talk later becomes an internalized thought pattern as the child encounters many instances of similar decision making.

One of the most effective techniques for managing groups of young children is teacher demonstration of desired behavior while explaining the reason that the modeled behavior fits the particular situation. The verbalization by the teacher of the adult reasoning behind expected behaviors provides to the young child additional information for understanding adult expectations and for organizing both thought and behavior in future situations.

Moral Values

A second major strategy in the management of young children originates from Kohlberg's (1975) stages of moral development. Kohlberg noted that the young child may be at a preconventional level: responsive to environmental rules and labels of good and bad or right and wrong, but interpreting such rules and labels in terms of the personal pain or pleasure gained from following the rules or doing what is good or right. Avoidance of pain or punishment and an unquestioning deference to authority figures who control the pleasureful or painful consequences of behavior are the primary values used by the preconventional child in selecting behaviors to exhibit. Teachers of young children who establish personal influence through deliberately associating themselves and desired behaviors with pleasureful consequences can utilize the child's natural values in classroom management.

Specific statements of expected behavior that will gain desired consequences from the teacher help young children avoid painful incidents and strengthen their trust in authority figures. The trust built at this early stage becomes the foundation for respect for the opinion of significant other people concerning what is right and wrong.

MANAGING THE ELEMENTARY MATHEMATICS CLASSROOM

The elementary mathematics teacher often finds that classroom management is most directly related to the effectiveness of the teacher in communicating personal influence to individual students, in building cohesive group monitoring of behavior, and in preventing problems by providing for individual differences in learning mathematics. These three management strategies permit the elementary teacher to maximize the use of personal influence (which passes its peak of effectiveness in the intermediate grades) while developing the peer group and self-directed management desirable for junior high or high school students.

Effectively Communicating Personal Influence

Effective communication between teacher and students is the foundation of good classroom management. When caring interpersonal interactions occur in the classroom the students' needs for safety, security, belongingness, and self-esteem are met. At the same time, the teacher meets personal and professional goals. Specific techniques for developing communication skills help the teacher strengthen personal influence in managing individual students.

"I" Messages

Gordon (1974) suggests that the first step in communication that meets mutual needs is the teacher's identification of "who owns the problem" when problem situations arise in the classroom. Teachers should ask themselves "Am I feeling unaccepting of the behavior because (1) I am being interfered with, damaged, hurt, impaired, or (2) because I want the student to act differently." Acknowledging that the first part of the question is true, as well as the second part, helps the teacher recognize that his or her personal feelings are part of the problem. Effective truthful communication to the student involves statements that include "I" messages of personal discomfort stemming from the problem behavior.

The components of an "I" message are:

1. Teacher states the problem behavior: "When I'm interrupted . . ."; "When you don't pick up. . . ."

2. Teacher states the personal effect of the behavior: "When I'm interrupted I lose my train of thought. . . ." "When you don't pick up I have to spend my time putting things away."
3. Teacher states the feelings this generates: "When I'm interrupted I lose my train of thought and feel anxious about regaining it." "When you don't pick up I have to spend my time putting things away and I feel imposed upon."

The teacher's use of "I" messages to describe to students their inappropriate behavior and its personal effect is a form of interpersonal communication that allows students to come to know the teacher at a person-to-person level. "I" messages model for students an effective way of assertively stating the personal feelings resulting from another person's behavior. When students copy the model in peer communication interpersonal problems are clearly stated and can be openly discussed. The message sender acknowledges responsibility for the feelings stated while placing responsibility for the initiating behavior at its source. Both teachers and students who state their feelings and needs through "I" messages strengthen the honesty and openness of classroom communication.

Communication Skills

Communication skills in the classroom include both sending and receiving. Jones and Jones (1981) provide the following suggestions (adapted with permission):

- Deal in the present—discuss important matters as soon as they occur when possible.
- Talk directly to the children rather than about them—even when other adults are present, children should be addressed directly, not discussed.
- Speak courteously—since teachers serve as models their interactions with children should include more courtesies than with adults.
- Make eye contact and be aware of nonverbal messages being sent—children often read the nonverbal messages as more important than the verbal content, e.g., looking past the child while praising work conveys lack of real interest or concern.
- Take responsibility for statements by using "I"—the teacher's skill in sending "I" messages helps students take responsibility for the effects of their behavior.
- Give specific, nonevaluative feedback—statements that clearly describe the inappropriate behavior and its effects are more precise in helping students change the behavior than value judgments (e.g., "When you visit with Ann

while I am discussing the new problems, I find it hard to keep the group's attention, and I feel very frustrated'' rather than ''How can you be so rude?'').

- Make statements rather than ask questions—misbehavior often results in a bombardment of teacher questions rather than ''I'' statements (e.g., ''I was getting concerned because you weren't here and I needed to begin explaining the new problems'' rather than ''Where have *you* been?'').
- Maintain a high ratio of positive to negative statements—teachers often are more responsive to inappropriate than appropriate behavior but student success and happiness in the classroom are the result of frequent messages from the teacher that they are valued, able, and responsible. (Jones & Jones, 1981, pp. 102–105)

Assertive Discipline

Effective teacher communication of personal influence is also dependent on the personal power of the teacher. Canter's (1976) approach to helping teachers increase their influence in the classroom stresses the need for teacher assertiveness. The assertive teacher is defined as:

> . . . one who clearly and firmly communicates her wants and needs to the students and is prepared to reinforce her words with appropriate actions. She responds in a manner which maximizes the potential to get her needs met, but in no way violates the best interests of the students. (Canter, 1976, p. 9)

Assertive teachers recognize their rights to establish classroom management procedures based on their personal preferences. Students are informed that they have the right to choose how to behave but that each type of problem behavior will have specifically defined consequences that will automatically follow.

The assertive teacher determines what behaviors are needed from students to enable the teacher to function at maximum potential. Although these behaviors may vary from activity to activity, they are clearly stated at the beginning of the lesson as the behavior expected by the teacher. If a student does not meet the stated expectations, the teacher may request the behavior by different methods:

1. hints—"Everyone should be filling out lab guidesheets.''
2. questions—"Have you filled out your lab guidesheet?''
3. ''I'' messages—"I want you to stop talking and fill out your lab guidesheet.''
4. demands—"Get out your lab guidesheet and fill it out.''

If the student refuses to respond to any of the request forms, an assertive confrontation between the teacher and the individual student may be necessary. In confrontation the teacher clearly, matter-of-factly, and firmly states only to the individual student the behavior expected, the consequences if the student chooses not to comply, and why this is being done. Calm, confident persistence in restating the expected behavior and consequences for noncompliance may occur as many as three times before the consequences are applied. When the student chooses to take the consequences rather than change the behavior, the teacher stresses the fact that the student has made the choice. The assertive teacher conveys to the student through statements and behavior the message:

> I care too much about you to let you disrupt, ruin your learning, and not be all you are capable of being. Therefore, I will set limits on your behavior and provide consequences if you choose not to stay within them.

Paraphrasing

In addition to being skilled in sending personal influence messages to students, the effective classroom manager is also skilled in listening. Nonevaluative listening provides students with a feeling of being clearly heard and accepted. In their description of guidelines for paraphrasing, i.e., active or reflective listening, Johnson and Johnson (1975) have stated several essentials of good listening skills. They include:

- restating the speaker's ideas and feelings in your own words to let the student know how you perceive what he or she is saying
- beginning your restatements with phrases such as "You are saying . . ." "You think . . ." "You feel that . . ." "It seems to you that . . ." or similar introductions to your paraphrase
- deliberately refraining from any indications of approval or disapproval of what the student is saying
- using nonverbal communication such as eye contact, facial expression, body language, and voice intonation to convey to the student your openness to both the ideas and feelings being expressed and your concentration on understanding what is being communicated
- stating as clearly and precisely as possible what you hear the student saying and the attitudes and feelings being conveyed
- refraining from manipulating or elaborating the student's message by adding to or subtracting from what is said
- imagining yourself in the student's place in order to better understand the meaning

Paraphrasing has the advantage of making the speaker aware of what the listener is hearing. Such statements are especially helpful to children because they often clarify how adult perceptions vary from the child's intent in communication. When the paraphrased message differs significantly from the intent of the message, the student has an opportunity to correct and clarify the original statement.

Reality Problem Solving

The elementary mathematics teacher who teaches students in the upper elementary grades quickly realizes that students are less responsive to direct personal influence as they become aware of the variety of behaviors acceptable in school classrooms. The teacher of older elementary students who recognizes this development of behavioral alternatives as an essential preliminary to self-directed behavioral control may find a problem-solving approach to classroom management more successful.

Glasser (1965, 1969) has described several step-by-step approaches for managing inappropriate behavior through problem solving. These sequential procedures have the advantages of being accomplished in a short period of time and of allowing the teacher to analyze and improve particular steps when the technique is not initially effective. Glasser's problem-solving approach is usually presented as a seven-step process:

- *Step One*—The teacher establishes a warm, supportive rapport with the student and consistently indicates willingness to be personally involved in the student's life.
- *Step Two*—The teacher asks the student to describe his or her current problematic behavior as specifically and objectively as possible. Teachers may assist by asking questions or offering to describe to the student what they observed. No value judgment is involved.
- *Step Three*—The teacher helps the student make a value judgment about the behavior by asking such questions as "Is the behavior helping you? Is it helping me? Is it helping the group?" If the student answers "yes" he or she is asked to describe how it helps or to list the payoffs and costs of the behavior.
- *Step Four*—When the student decides the behavior needs to be changed, the teacher assists him or her in developing a workable plan for making the change. Students encouraged to devise their own plans become more self-directive but the teacher may suggest several alternatives if the student is unable to do so.
- *Step Five*—The teacher and student clarify the plan and the teacher asks the student to make a commitment to it. This final commitment to try the plan is an essential step.

- *Step Six*—The teacher follows up on the plan by setting aside a definite time to meet with the student to discuss how the plan is working and to reinforce the student for carrying it out.
- *Step Seven*—If the plan does not change the behavior, the first three steps are reviewed by the teacher and the student is asked to give an explanation of why the plan has not worked. No criticism or judgment is made by the teacher, but student excuses that blame failure on others are not accepted. The student is then assisted in devising another plan and making a new commitment.

Building Cohesive Group Monitoring of Behavior

Classroom management that utilizes group influence to control individual behavior promotes student development of social awareness and frees the teacher from overuse of personal influence. The elementary classroom provides an excellent age group and situation for teaching group problem-solving interactions and procedures.

Distinct procedures for classroom group problem solving have been delineated by Glasser (1969). Regular class meetings for the purpose of discussing classroom management allow teachers to acknowledge their respect for students' ability to work out their own problems. The active participation of students in discussing problems openly in a group setting helps them develop the kinds of procedures and statements that are most efficient. Teacher reinforcement of appropriate group participation also increases the likelihood that reticent students will try to exhibit similar behaviors in classroom meetings.

The problem-solving classroom meeting focuses on real problems that affect the entire group. Through group discussion the problem is clarified, additional information presented, alternative solutions proposed, and group commitment made to a plan for group problem-solving action. When the problem is the result of the behavior of an individual student the group both confronts the student with the effects of the behavior and proposes a plan for helping the student change the behavior. The misbehaving student has the opportunity to explain the effects of other students on his or her behavior and to suggest measures that will facilitate personal change.

The role of the teacher in the classroom meeting is to guide and facilitate purposeful communication and to participate as an individual member of the group. Procedures for guiding the meeting follow the seven-step plan for reality problem solving listed in the preceding section. The purpose of the meeting is twofold: helping students develop awareness of the effects of their behavior on others and developing students' problem-solving skills in proposing and implementing workable solutions.

Although the mathematics class focuses on content to be learned and skills to be acquired, the student in the classroom must learn to function as an effective group member to be prepared for the vocational and social responsibilities of adulthood. Classroom meetings emphasize the importance of social interaction and environmental feedback in solving problems. Such experiences help students understand that problem solutions vary with situations and individual perceptions, a mathematical truth as well as a social truth.

Preventing Problems by Providing for Individual Differences

The mathematics classroom in the elementary school has the flexibility that permits the teacher to prevent many problem behaviors. From an ecological perspective (Laten & Katz, 1975) problem behavior is often the result of a "poor fit" between the student's competence to do the assigned task, accompanied by the student's selection of behavior for the situation, and the teacher's expectations for competence and behavior, the nature and structure of the task itself, and the contingencies for reinforcement in the situation.

Teacher Expectations

Teacher expectations are sometimes negatively translated into instructional differences. Brophy and Good (1974) have collected research indicating the following differences (abridged by Rich, 1982) in teacher instruction of high- and low-achieving students:

1. Waiting Less Time for Lows To Answer: Teachers have been observed to provide more time for high achieving students to respond. . . .
2. Staying with Lows in Failure Situations: Teachers have been found to respond to lows' . . . incorrect answers by giving them the answer or calling on another student. . . .
3. Rewarding Inappropriate Behavior of Lows: Teachers have been found to praise marginal or inaccurate student responses. . . .
4. Criticizing Lows More Frequently Than Highs: Teachers have been found to criticize lows more frequently than highs when they provide wrong answers.
5. Praising Lows Less Frequently Than Highs: When lows provide correct answers they are less likely to be praised. . . .
6. Not Giving Feedback to Public Responses of Lows: Teachers . . . have been found to respond to lows' answers (especially incorrect answers) by calling on another student to respond. . . .
7. Paying Less Attention to Lows: Teachers attend more closely to highs . . . [e.g.,] smile more often and maintain greater eye contact. . . .

8. Calling on Lows Less Often: Teachers have been found to call on high achieving students more frequently. . . .
9. Differing Interaction Patterns of Highs and Lows: In elementary classrooms, highs dominate public response opportunities. . . . In secondary classrooms highs become even more dominant. . . .
10. Seating Lows Farther from the Teacher: Seating pattern studies have sometimes found that lows tend to be placed away from the teacher. . . .
11. Demanding Less from Lows: This is a broader concept suggesting such activities as giving these students easier tests (and letting the students know it) or simply not asking the student to do academic work. . . . (Rich, 1982, p. 220)

Structuring classrooms to fit student differences in competence for the task and in perception of expected behaviors simplifies effective classroom management.

The Engineered Classroom

The engineered classroom developed by Hewett (1968) is a precisely described system of classroom management that is teacher directed but provides for individual differences in readiness for learning and school-appropriate behavior. Hewett suggests that there is a developmental sequence for successfully reaching educational goals in any classroom. The sequence follows the order: (1) *attention*—paying attention to the relevant cues in the environment; (2) *response*—responding to appealing tasks; (3) *order*—responding in an orderly, direction-following manner; (4) *exploration*—exploring through sensory experiences, i.e., handling objects, doing experiments, seeing films, etc., the many sources of environmental information; (5) *social*—gaining social approval and avoiding disapproval; (6) *mastery*—mastering the basic skills needed for the curriculum content and the classroom; (7) *achievement*—being self-motivated and independent in pursuing learning.

The physical environment of the engineered classroom features learning centers with activities specifically designed to develop attention, response, order, exploration, and social interaction, as well as the usual classroom arrangement of desks for mastery and achievement activities. This classroom arrangement, depicted in Figure 10–1, permits the teacher to move students from one type of activity to another when problem behavior indicates that a student needs a less demanding task or a more personally motivating task.

Regular academic curriculum is done in the Mastery Center, teaching station, and study booths. Students at achievement level may be given an enriched curriculum designed to develop independence and creativity through such activities as special projects, developing demonstrations for other students, or

Figure 10–1 The Engineered Classroom

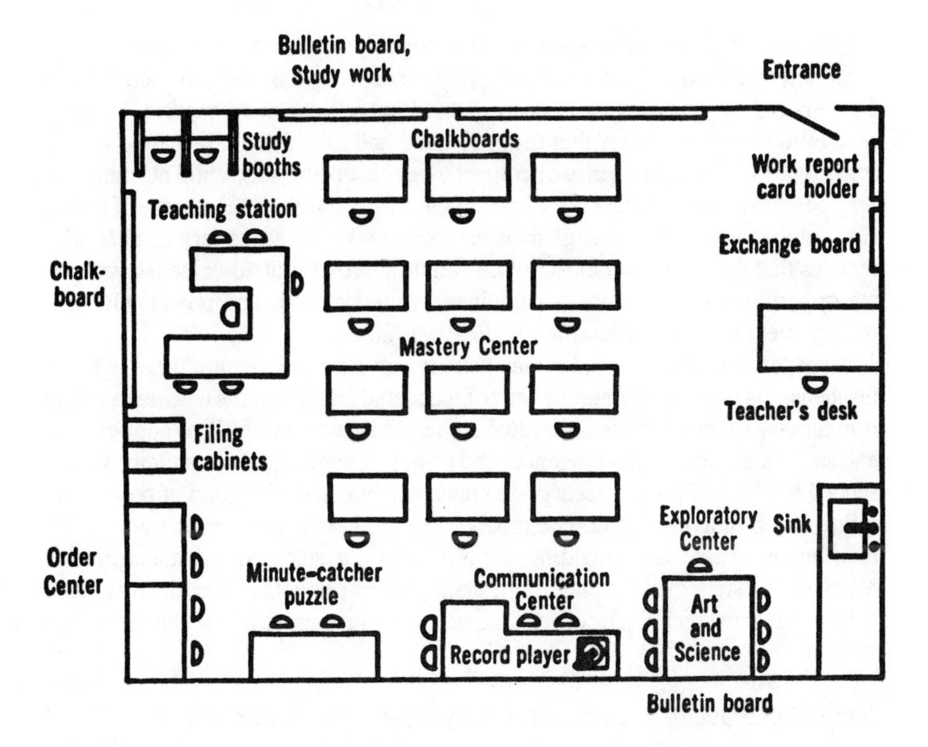

Source: Hewett, F.M., & Taylor, F.D. *The Emotionally Disturbed Child in the Classroom* [2nd ed.]. (1980). Boston: Allyn & Bacon, p. 216. Used by permission.

---

self-directed activities to increase speed of response in computation. Students at mastery level may be working on mathematical activities designed to bring their skills up to an expected performance criterion.

The Communication Center features mathematical games, group projects, or group experiments to help students increase social skills such as contributing to group work or waiting one's turn. The Exploratory Center functions as a minilab with a variety of "hands-on" experiences for actively exploring the mathematics of real life. The Order Center can contain simple mathematical tasks designed to help students learn to attend, follow directions, or complete assignments. Puzzles, mazes, coding, design copying, tracing, matching, geoboards, or other simple construction tasks may be appropriate. A work report-card holder by the door holds daily work assignments for each student and an exchange board tells how students can exchange completed work for rewards (Hewett & Taylor, 1980).

MANAGING THE SECONDARY MATHEMATICS CLASSROOM

Effective classroom management of secondary mathematics groups requires some understanding of adolescents' unique needs. In his study of self-esteem Coopersmith (1967) identified three variables that relate to these needs: (1) a sense of significance (i.e., belief that they are liked and important to someone who is important to them); (2) a sense of competence (i.e., being successful at some task they personally value that is reinforced in the classroom); and (3) a sense of power (i.e., ability to control parts of their environment). The secondary teacher who realizes that the classroom must provide adult-approved but adolescent-appropriate opportunities for obtaining significance, competence, and power will plan management strategies accordingly (Jones, 1980).

Management of the secondary mathematics groups may require three types of teacher skills to meet the unique needs of older students. First, the teacher may find it necessary to increase the individual student's awareness of his or her personal and social significance, competence, and power through direct communication. A second teacher skill that affects good management is the utilization of peer group influence on individual group members. The third skill, the eventual goal of all management, involves providing opportunities for self-directed management. Many of the strategies for classroom management previously described are also useful, but some techniques of management are more applicable to the secondary student.

Increasing Student Awareness through Direct Communication

As students approach adulthood, the selection of situationally appropriate behavior becomes more related to personal needs than to the rules and regulations of adult authority figures. Frequently, the secondary classroom teacher may encounter surface behaviors that reflect the adolescent's developing autonomy but are not indicative of a more pervasive problem. Through direct communication the teacher can increase the student's awareness of personal competence to choose more acceptable behaviors. Techniques for managing surface behaviors, clarifying expectations and procedures, therapeutic stress management, and life space interviewing help the secondary mathematics teacher communicate to older students their personal significance and their competence and power to change behaviors.

Managing Surface Behaviors

Several possible intervention techniques may be used in managing surface behaviors to reduce the risk of escalation or contagion of individual problem behaviors (Redl & Wineman, 1957). These techniques are particularly useful in making the older student aware that the teacher recognizes the behavior as indicative of a need for significance, competence, or power:

- *Planned ignoring*. Recognizing that the student's behavior is being used for its "goading or power game playing" effect and will probably dissipate if left unchallenged.
- *Signal interference*. Using nonverbal signals such as facial expressions, hand movements, or body language to let the student know that a behavior is inappropriate. A signal indicates the teacher feels the student is competent to change the behavior without teacher direction.
- *Proximity and touch control*. Moving near or standing next to the student, pointing to the assigned task, placing a hand on the shoulder to indicate that the student is significant to the teacher and there is concern for the cause of the present behavior.
- *Involvement in interest relationships*. Engaging the student in a personal conversation about some topic of interest to indicate the significance of the student's personal interests and their relationship to the classroom activities.
- *"Hypodermic" application*. Communicating to the student genuine feelings of affection to increase the student's sense of personal significance.
- *Tension decontamination through humor*. Using humor, affectionate kidding, or an appropriate joke to divert attention and allow a student a face-saving out when tension is building.
- *Hurdle help*. Providing help with the assigned task as a means of decreasing student anxiety and frustration concerning competency to do what is expected.
- *Interpretation as interference*. Taking time to discuss with the student some classroom situation, teacher statement, or peer action that has been misinterpreted by the student and is causing problem behavior.
- *Regrouping*. Making changes in work groups to reduce growing tension and friction among the group members.
- *Restructuring*. Abandoning a planned or routinely scheduled activity when it is not well received by the group and letting the students "brainstorm" other ways to learn the same information or the relevance of the learning to their personal lives.
- *Direct appeal*. Using teacher personal influence in an honest appeal (e.g., "Hey, that's enough! I need to get you through this work so you can pass the state competency exam") to stop problem behavior.
- *Limiting space and tools*. Diverting students' attention from seductive objects in the environment (which range from minicalculators to good-

looking classmates) by separating the student from the object that is so distracting that the student cannot handle the situation alone.

- *Antiseptic bouncing*. When a student has reached the point that other techniques are not useful, removing that student from the group before there is total loss of self-control, thereby providing an opportunity to "cool off" or regain control.

Clarifying Expectations and Procedures

Problem behavior in the secondary mathematics classroom may result from a student's lack of understanding of teacher expectations. Students who are frequently disruptive seem unable to recognize and utilize the available opportunities for increasing their significance, competence, or power and resort to other means. Poor perception of what is expected and its relationship to meeting adolescent needs may be aided by Jones' Model for Clarifying Expectations shown below. This model provides a means for teacher clarification of problems, and for feedback to students on specific unacceptable behaviors. It also provides the student with a means of communicating personal needs, personal feelings about teacher expectations, and willingness to negotiate differences. This approach to problem solving is specifically designed to enhance a student's feelings that he or she is significant to the teacher, has competencies that can be applied in the classroom, and has the power of negotiation with another adult. Jones' model is specifically suggested for mainstreaming or readmitting behavior-problem adolescents.

Step 1: Teacher lists the problems as she perceives them
- problems should be stated in terms of specific observable behaviors
- if possible data should be collected
  1. data help determine if the behavior requires intervention
  2. data provide a base from which to determine if an intervention is effective

Step 2: Teacher lists expectations
- stated positively
- include 1–5 goals
- goals are observable and measurable
- goals include the minimum level of acceptable behavior

Step 3: Teacher lists what she is willing to do in order to enhance the likelihood that the expectations can be met

Step 4: Student lists the problems as he perceives them
- problems should be stated in terms of specific observable behaviors

Step 5: Student examines the teacher's list of problems and determines what behaviors do in fact occur and create a problem

Step 6: Student examines the teacher's list of expectations and indicates those which he can and cannot meet

Step 7: The student lists what he needs in the class in order to meet the teacher's expectations

- needs must be stated in terms of specific observable behaviors
- stated positively
- include 1–5 needs
- include the minimum level of acceptable behavior

Step 8: Student lists what he is willing to do in order to enhance the likelihood that the teacher's expectations can be met

Step 9: Teacher examines the student's statements concerning:

- which teacher expectations can be met
- what needs the student has
- what the student is willing to do

Step 10: Student and teacher meet and develop a plan which states each party's responsibilities

- responsibilities are listed in terms of specific observable behaviors
- responsibilities are stated positively
- responsibilities include a minimum level of acceptable behavior

Step 11: Teacher and student develop a safety net plan

- plan lists what each party will do if he or she believes the other has violated the agreement

Step 12: Evaluation

- includes periodic conferences to examine the program's progress
- includes specific data collection when possible (Jones, 1980, p. 206)

Therapeutic Management

Often the cause of problem behavior in the secondary classroom is excessive stress related to the complexities and frequently rapid changes of modern society. Pupils under stress may react to their feelings with either aggressive or withdrawn behaviors and this may have a contagion effect on the teacher. Long and Fagen (1981) suggest a number of teaching strategies to help students cope with stress:

- *Forming a helping adult relationship*. By accepting problems matter-of-factly and being sympathetic while insisting that the student take some responsibility for improving the stressful situation and reinforcing attempts to do so, the teacher focuses attention on how to make the situation better.
- *Lowering school pressure*. When stress seems excessive, the teacher can temporarily lower academic requirements and extend deadlines.
- *Redirecting feelings into acceptable behavior*. While accepting the presence of negative feelings, the teacher provides socially approved behavioral alternatives such as construction tasks that utilize physical energy productively.
- *Learning to accept disappointment and failure*. Making students aware of their unrealistic self-expectations that they should always be successful and that everyone should like them lessens the stress of human relationships and competitiveness.
- *Completing one task at a time*. A helpful teacher can work with a student to reduce a seemingly unmanageable number of tasks to a manageable unit of work.
- *Helping other students*. By helping other students whose problems are more severe or who are younger, students see their own problems as less stressful.
- *Separating from the setting*. Occasionally the best coping strategy is to leave a stressful environment temporarily. Teacher interaction may be needed to clear the way for such action.
- *Seeking professional help*. When stress appears to be chronic and overwhelming, the classroom teacher should refer the student to the counselor or other appropriate personnel. (Adapted from Long & Fagen, 1981, pp. 167–172)

The Life Space Interview

A particularly effective technique for teacher communication aimed at increasing students' awareness of their personal perceptions is the Life Space Interview (Morse, 1963; Redl, 1959). Through a series of questions asked by the teacher in the immediate context of the problem behavior the student is aided in clarifying the problem and finding a solution. Several interview questions can be used:

1. In a nonthreatening manner ask the student or students to tell "What happened?" in order to get their perceptions.
2. By paraphrasing the students' responses, try to clarify whether the presenting problem is the issue or whether there are other related problems that have distorted the students' perception of the incident.
3. Ask the student "How do you feel about all this?"
4. Ask the student "What do you think should be done about this?"

5. Discuss the realities of what may happen should the behavior continue.
6. Discuss with the student how he or she might be helped in controlling the behavior.
7. Develop a plan for alternative behaviors to be used should the problem situation arise again.

The Life Space Interview allows the teacher to facilitate student competence in defining real-life problems and student power in selecting alternative solutions.

Utilizing Peer Group Power

A major developmental task of adolescence is establishing healthy, mutually satisfying peer relationships (Jones, 1980). Concern for these relationships frequently overshadows any academic interests. The classroom teacher who is skilled in using peer group cohesiveness as a behavior control strategy seldom has major problems in secondary classroom management. Two management techniques that emphasize peer group acceptance are Adler's and Dreikurs' and Cassel's social theory and Jones' peer feedback.

Social Motivation

A particularly useful theory in understanding adolescent classroom behavior is Alfred Adler's belief that the central motivation for all human behavior is to belong, to be accepted, and to be fulfilled within a group setting (Krasnow, 1971). Disturbing behavior is viewed as the result of discouragement or insecurity concerning group acceptance. Such behavior is motivated by four faulty goals for gaining acceptance: (1) attention getting as a means of being recognized by the group; (2) power and control, particularly as a means to defeat the power of authority figures and thus gain group status; (3) revenge as a means of getting even for what others (teachers or students) have done to hamper group acceptance; and (4) helplessness or inadequacy as an indication of giving up on the possibility of group acceptance or status. The classroom teacher's role in managing students motivated by these faulty goals is twofold: first, to identify which of the goals the student is seeking and verify it by teacher feelings resulting from the behavior, and, second, to help the student recognize that the faulty goals do not gain the group acceptance sought.

A four-step process is suggested by Wolfgang and Glickman (1980) to identify and change faulty goals that motivate problem behavior. First there is a period of observation and data collection to record the student's behavior. As a second step the teacher hypothesizes which of the underlying goals is the source of the behavior. In the third step the hypothesis is verified by the teacher's reflection on personal feelings resulting from the behavior: (1) if the teacher's feeling was annoyance, the student's goal was probably attention getting; (2) if the teacher's feeling was one of being intimidated or beaten, the student's goal was probably power or control; (3) if the teacher's feeling was one of being wronged or hurt, the student's goal was probably revenge; and (4) if the teacher felt incapable of reaching the student, the goal was probably helplessness.

The final step in the process is to confront the student (after an episode of disruptive behavior) with the hypothesized goal. The confrontation may begin with the teacher's question "Do you know why you acted as you did today?" followed by one of the four possible goals: (1) "Could it be you want special attention?" (2) "Could it be you want your own way and hope to be boss?" (3) "Could it be you want to hurt others as much as you feel hurt by them?" or (4) "Could it be you want to be left alone?" The student's response is the final verification of the motivation prompting the behavior (Adapted from Wolfgang & Glickman, 1980, p. 83).

Once the goal has been defined, the teacher interventions suggested by Dreikurs and Cassel (1972) may be employed. Students who seek attention receive deliberately planned teacher and group attention for productive group behavior. When the student who seeks power tries to engage the teacher in a confrontation, the teacher removes the issue by stating that he or she cannot force compliance and suggests that the student make his or her own choice regarding nondisruptive behavior so that the group can continue. In addition to refusing power struggles, the teacher may specifically design activities that will enhance the student's status within the group (e.g., have the student plan a mathematical experiment and demonstrate it for the group). For the student who seeks revenge, the teacher may plan specific personal activities or peer activities to convince the student that he or she is personally liked and valued and has no real cause for seeking to get even. Students who have resorted to the goal of helplessness are the most difficult to deal with. They require teacher patience and flexibility in finding classroom activities that the student will even try. Praise for attempting tasks and provision of peer-cooperative activities may be useful in encouraging the student who has given up on ever being a successful classroom group participant.

Peer Feedback

An extremely effective but potentially problematic strategy for managing classroom behavior of adolescents is peer feedback (Jones, 1980). If properly structured and facilitated by the teacher, peer feedback can change behaviors when other strategies have failed.

Initially the problem student is asked if he or she would like to participate in a classroom group discussion of his or her disruptive behavior. If the student chooses not to be present, a videotaping of the group discussion is suggested so that the student may later see and respond to the group's statements.

The discussion begins with informing the group that there is concern about the disruptive behavior of Student X and asking if they will be willing to work out a plan to help Student X. Discussion is then directed toward defining the behaviors that disrupt, identifying possible causes (e.g., group reinforcement for attention-getting behavior in a dull class), and making specific suggestions for group behaviors to aid Student X. Suggestions may include (1) ignoring attention-getting behaviors; (2) increasing positive interactions with Student X outside the classroom, such as talking in hallway, offering help with assignments, etc.; and (3) allowing class members to ask that Student X be removed from the classroom when disruptive behavior is excessive and using group majority vote as the decision for removal.

Since peer group feedback is often emotionally laden, the teacher may not choose to involve the entire classroom group. A more productive procedure may be to identify a small group of students within the larger group who will meet once or twice a week to help a problem student make a better group adjustment. Opportunities for both positive and negative feedback can be planned in the group's interactions with the problem student. Negative feedback is structured to focus on specific behavior, to deal only with behavior the individual can change, and, when possible, to provide alternative behaviors that would be more peer acceptable. The teacher serves as a model for appropriately stating to the problem student both his or her behaviors that result in positive regard and those that trigger negative feelings in other students.

Providing Opportunities for Self-Directed Management

The secondary mathematics classroom, whether structured for direct instruction or for laboratory experiments, provides an excellent opportunity for the teacher to facilitate or actively instruct students in self-management. Techniques for self-management such as self-monitoring, contingency contracting, and cognitive behavior modification may initially require some teacher direction until the student adequately understands and can utilize the procedures across varied tasks and settings. Eventually, this translates to self-confidence regarding competence to select appropriate behavior.

Self-Monitoring

Adolescents growing up in a complex, changing society often experience confusion and stress in comprehending the multiple value systems espoused by adults, or even peers. When confusion in value selection produces enough discomfort, the young person may resort to various defense mechanisms (Jones, 1980, p. 166). Such defense mechanisms as rationalization, repression, or aggression often prevent the adolescent from developing a realistic awareness of his or her own behavior and its effects. Increased awareness of personal behavior and its result is the basis for self-directed management.

Self-monitoring of specific behaviors is a useful intervention strategy when student motivation and impulse control are sufficient (Workman, 1982). Simply tallying the occurrence of behaviors that are disruptive frequently brings about a change. Tallying may be done by countoon (Jones, 1980, p. 168), hand counters, or paper and pencil. Specific behavioral counts may also make the teacher aware that the behavior is not occurring as frequently as thought. If verification of the count is desirable, a second observer can be used to record the behavioral occurrences. Comparison of the student's self-report and the report of the external observer verifies the student's competence in self-reporting.

Student monitoring of behavior may also be extremely useful in bringing about changes in interpersonal interactions between student and student or between teacher and student. Tallying the number of positive and/or negative statements made to others helps the student become more aware of the effects of each type of statement. Likewise, counting the positive and negative statements of peers in particular environments or situations may clarify for the student relationships between his or her actions and peer statements.

Collecting counts of the teacher's positive and negative comments may serve as the first step in teaching students to monitor teacher feedback (Gray, Graubard, & Rosenberg, 1974). After collecting such data, students can be taught methods for reinforcing positive contacts with the teacher, including eye contact, pleasant approving facial expression, expressing gratitude for personal help and attention, expressing excitement over the ideas and content being taught, and tactfully letting the teacher know when teacher statements or behavior have not been useful. A second count of teacher behavior and comments after the student has been applying these methods for reinforcing positive contacts will usually help the student become aware of his or her own power to increase positive interactions with others.

A second method of increasing student awareness of a wide range of classroom behaviors is videotaping or audiotaping. Providing adolescents with an opportunity to observe their behaviors or hear their language interactions in an undeniable, nonmemory-based format allows them to compare their initial perceptions with a second view or hearing of what actually occurred. If the situation is emotionally charged, this review may be postponed until a later date when feelings have dissipated. Productive videotaping as a means of increasing student awareness requires that the taping process be so accepted that it no longer produces behavior change in and of itself. When taping is an ongoing process in the classroom it is possible to demonstrate to students both their productive and nonproductive behaviors and instances of significance, competence, and power.

Another technique used to initiate student self-direction of behavior is titled Stop Action (Jones, 1980, p. 177). In this strategy students agree to stop or freeze immediately upon hearing a designated cue. The sole intent of the activity is to increase student awareness and no negative consequences should be associated with the procedure. Two approaches may be used to monitor the behavior occurring when the "Stop Action" cue was given: first, students may be asked simply to record what they were doing each time the signal was given; second, students may designate a specific behavior they wish to either decrease or increase, describe or draw a picture of the behavior and place it on their desk, and, on the cue, mark whether or not the behavior was occurring.

A potentially useful technique for the mathematics classroom is having students write a log or journal describing environmental stress factors and their own behavioral responses. This technique can provide many positive benefits: (1) a chance for the student to relax in self-expression apart from task and teacher expectations; (2) an opportunity to reflect on what has happened and analyze it in a language form; (3) an effective tool for interpersonal crisis situations, allowing participants to privately clarify the event and then compare results as a first step toward conflict resolution; and (4) a chance for the adolescent student to slow down, examine problems, and be self-directive in proposing alternative behaviors or solutions.

Providing opportunities for students to develop self-monitoring skills lays a foundation for other forms of self-management. The time spent in teaching these skills may have more lifetime influence than much of the content taught in secondary mathematics.

Contingency Contracting

Although the use of contracts is usually viewed as a teacher-directed management strategy, Homme's (1970) five levels of contingency contracting should be included here. Levels progress from manager-controlled (i.e., teacher contracts) to student-controlled contracts with three transitional levels in between as the student assumes more responsibility for deciding the reinforcement to be given and determining the amount of the task to be done. To aid the student in reaching the final stage of contractual self-management, it is suggested that microcontracts for specific tasks be combined into macrocontracts for the completion of a set number of microcontracts. The ultimate goal of contingency contracting is to teach students to both establish and fulfill their own contracts and to reinforce themselves, under macrocontracts, for such fulfillments.

Cognitive-Behavioral Intervention

In an interesting new approach to helping students develop internal self-management strategies, Meichenbaum (1982) utilizes external language to train internal thoughts. Cognitive-behavioral interventions teach students to talk aloud to themselves as a means of ensuring that they think before they act. Students are taught a series of self-statements or questions to be asked and answered in approaching either social or academic problems.

The premise that socialization is a process of hearing significant others give instructions, repeating these instructions aloud, then internalizing the instructions into thought patterns is basic to the training process. Students watch and listen as the teacher models a procedure for self-instruction and questioning such as the following:

> I must stop and think about this before I act. What plans can I try? What would happen if I tried that? That's not right. It isn't working. What else could I try next? That seems to be working. Good for me! I worked it out for myself!

The questions may vary with the content of the problem situation, but the strategy models for the student the self-monitoring, error correction, self-evaluation, and self-reinforcement of good logical problem-solving thinking.

After the teacher models the procedure the student applies the same self-questioning and responding process by talking aloud to himself or herself in a problem-solving situation. The overt rehearsal allows the teacher to monitor the student's use of the technique. Additional whispered rehearsals during regular class time may be desirable as the student tries the procedure in various situations. After sufficient rehearsal the student is encouraged to "think" the procedure without the externalization of the words. The procedure may revert to overt language if the student or teacher desires further monitoring of the process.

SUMMARY

The "orchestration" of the mathematics classroom is a process of effective interpersonal communication, productive group utilization, and planning for individual differences. At every age from preschool to secondary the teacher serves as a model for communicating personal feelings and expectations so that interpersonal relationships do not result in problems. As students grow older management strategies provide many opportunities to teach students effective group problem solving and self-management. Facilitating the learning of both the content of the mathematics curriculum and the principles of effective human relations and self-direction is the goal of good classroom management.

REFERENCES

Aronson, E. (1978). *The jigsaw classroom*. Beverly Hills, Calif.: Sage.

Bower, E. (1969). *Early identification of emotionally handicapped children in school*. Springfield, Ill.: Charles C Thomas.

Brophy, J.E., & Good, T.L. (1974). *Teacher-student relationships: Causes and consequences.* New York: Holt, Rinehart & Winston.

Brophy, J.E., Good, T.L., & Nedler, S.E. (1975). *Teaching in the preschool.* New York: Harper & Row.

Bruner, J.S. (1963). *Toward a theory of instruction.* Cambridge, Mass.: Harvard University Press.

Canter, L. (1976). *Assertive discipline: A take charge approach for today's educator.* Los Angeles: Canter.

Case, R. (1978). Developmental theory of instruction. *Review of Educational Research, 48,* 439–463.

Coopersmith, S. (1967). *The antecedents of self-esteem.* San Francisco: W.H. Freeman.

DeLoach, T.E., Earl, J.M., Brown, B.S., Poplin, M.S., & Warner, M.M. (1981). LD teachers perception of severely learning disabled students. *Learning Disability Quarterly, 4,* 343–358.

Dreikurs, R., & Cassel, P. (1972). *Discipline without tears: What to do with children who misbehave.* New York: Hawthorn Books.

Dunn, R., & Dunn, K. (1972). *Practical approaches to individualizing instruction: Contracts and other effective teaching strategies.* West Nyack, N.Y.: Parker.

Gallagher, J.M., & Reid, D.K. (1981). *The learning theory of Piaget and Inhelder.* Monterey, Calif.: Brooks/Cole.

Getzels, J.W., & Thelen, H.A. (1960). The classroom group as a unique social system. In N. Henry (Ed.), *The dynamics of instructional groups* (pp. 53–82). (Fifty-ninth Yearbook of the National Society for the Study of Education, Part II). Chicago: The University of Chicago Press.

Glasser, W. (1965). *Reality therapy.* New York: Harper & Row.

Glasser, W. (1969). *Schools without failure.* New York: Harper & Row.

Gordon, T. (1974). *Teacher effectiveness training.* New York: David McKay.

Gray, R., Graubard, P., & Rosenberg, H. (1974, March). Little brother is changing you. *Psychology Today,* pp. 42–46.

Hamrick, K.B. (1980). Are you introducing mathematical symbols too soon? *Arithmetic Teacher, 23*(3), 14–15.

Hewett, F.M. (1968). *The emotionally disturbed child in the classroom.* Boston: Allyn & Bacon.

Hewett, F.M., & Taylor, F.D. (1980). *The emotionally disturbed child in the classroom* (2nd ed.). Boston: Allyn & Bacon.

Hiebert, J., Carpenter, T.P., & Moser, J.M. (1982). Cognitive development and children's solutions to verbal arithmetic problems. *Journal for Research in Mathematics Education, 13,* 83–98.

Homme, L. (1970). *How to use contingency contracting in the classroom.* Champaign, Ill.: Research Press.

Johnson, D., & Johnson, R. (1975). *Learning together and alone: Cooperation, competition and individualization.* Englewood Cliffs, N.J.: Prentice-Hall.

Johnson, D.A., & Rising, G.R. (1972). *Guidelines for teaching mathematics* (2nd ed.). Belmont, Calif.: Wadsworth.

Johnson, M.L., & Wilson, J.W. (1976). Mathematics. In C. Seefeldt (Ed.), *Curriculum for the preschool-primary child: A review of research.* Columbus, Ohio: Charles C. Merrill.

Jones, V.F. (1980). *Adolescents with behavior problems: Strategies for teaching, counseling and parent involvement.* Boston: Allyn & Bacon.

Jones, V.F., & Jones, L.S. (1981). *Responsible classroom discipline.* Boston: Allyn & Bacon.

Kagan, J., Moss, H.A., & Sigel, I.E. (1963). Psychological significance of styles of conceptualization. In J.C. Wright & J. Kagan (Eds.), Basic cognitive processes in children (pp. 73–112). *Monographs of the Society for Research in Child Development, 28*(2 Serial No. 86).

Kagan, J., Rosman, B.L., Day, D., Albert, J., & Phillips, W. (1964). Information processing in the child: Significance of analytic and reflective attitudes. *Psychological Monographs, 78*(1, Whole No. 578).

Kidd, K., Myers, S., & Cilley, D. (1970). *The laboratory approach to mathematics*. Chicago: Science Research Associates.

Kohlberg, L. (1975). The cognitive-developmental approach to moral education. *Phi Delta Kappan, 56*, 670–677.

Krasnow, A. (1971). An Adlerian approach to the problem of school maladjustment. *Academic Therapy Quarterly, 7*, 171–183.

Laten, S., & Katz, G. (1975). *A theoretical model for assessment of adolescents: The ecological/behavioral approach*. Madison, Wis.: Madison Public Schools, Special Educational Services.

Lemlech, J.K. (1979). *Classroom management*. New York: Harper & Row.

Long, N.J., & Fagen, S.A. (1980). Therapeutic management: A psychoeducational approach. In G. Brown, R.L. McDowell, & J. Smith (Eds.), *Educating adolescents with behavior disorders* (pp. 159–175). Columbus, Ohio: Charles E. Merrill.

Meichenbaum, D. (1982). *Teaching thinking: A cognitive-behavioral approach*. Austin, Tex.: Society for Learning Disabilities and Remedial Education.

Morse, W.C. (1963). Working papers: Training teachers in life space interviewing. *American Journal of Orthopsychiatry, 33*, 727–730.

Pine, G.J., & Boy, A.V. (1977). *Learner centered teaching: A humanistic view*. Denver: Love Publishing.

Redl, F. (1959). The life space interview. *American Journal of Orthopsychiatry, 29*, 5.

Redl, F., & Wineman, D. (1957). *The aggressive child*. New York: The Free Press.

Rice, B. (1979, September). Brave new world of intelligence testing. *Psychology Today*, pp. 27–41.

Rich, H.L. (1982). *Disturbed students: Characteristics and educational strategies*. Baltimore: University Park Press.

Sharma, M.C. (1979). Levels of knowing mathematics. *Mathematics education monograph series, 1*(3). Framingham, Mass.: The Center for Teaching/Learning of Mathematics.

Sigel, I.E., & Coop, R.H. (1974). Cognitive style and classroom practice. In K.H. Coop & K. White (Eds.), *Psychological concepts in the classroom* (pp. 250–275). New York: Harper & Row.

Silbert, J., Carnine, D., & Stein, M. (1981). *Direct instruction mathematics*. Columbus, Ohio: Charles E. Merrill.

Slavin, R.E. (1980). Cooperative learning. *Review of Educational Research, 59*, 314–342.

Slavin, R.E. (1981). A policy choice: Cooperative or competitive learning. *Character, 2*(3), 1–6.

Souviney, R.J. (1980). Cognitive competence and mathematical development. *Journal of Research in Mathematics Education, 11*, 215–224.

Vygotsky, L.S. (1977). *Thought and language*. Cambridge, Mass.: MIT Press.

Wallen, C.J., & Wallen, L.L. (1978). *Effective classroom management* (abridged ed.). Boston: Allyn & Bacon.

Witkin, H.A., Moore, C.A., Goodenough, D.R., & Cox, W. (1977). Field-dependent and field-independent cognitive styles and their educational implications. *Review of Educational Research, 47*, 1–64.

Wolfgang, C.H., & Glickman, C.D (1980). *Solving discipline problems: Strategies for classroom teachers*. Boston: Allyn & Bacon.

Workman, E.A. (1982). *Teaching behavioral self-control to students*. Austin, Tex.: Pro-Ed.

Chapter 11

The Effective Use of Divergent Delivery Systems with Learning-Disabled Students

Raymond E. Webster

The terms *least restrictive environment* and *mainstreaming* have become popular to describe educational placements for learning-disabled students. In daily practice the two terms are often used interchangeably to mean that all learning-disabled students must be educated in the regular classroom setting, regardless of their individual educational needs.

Least restrictive environment and *mainstreaming* are not equivalent terms. In fact, each means something very different from the other. Mainstreaming involves integrating the student with special needs within the regular classroom setting. This placement may or may not be the least restrictive environment for that student. The least restrictive environment is that educational setting which is most closely like the rest of the educational setting where the student can derive maximum benefit from the school program when one considers the student's unique educational and social-emotional needs.

Some learning-disabled students may require only one or two hours weekly of help from a resource teacher to organize class assignments in mathematics and to discuss specific problems in understanding certain concepts such as irrational numbers or theorems. The least restrictive environment for such a student would be 28 hours a week in regular education and two hours weekly in special education. More severely learning-disabled students may require intensive special education in all academic areas and spend 21 hours weekly in special education. The remaining 9 hours of school time would be spent mainstreamed in art, music, and physical education.

In general, mainstreamed students are those with mild learning and/or behavior problems. To a lesser extent students with moderately severe problems in either of these areas may be mainstreamed for part of their school day. These students spend a large proportion of the school day in a self-contained special class. This educational placement is the least restrictive environment that best meets their unique needs. Severely learning-disabled students might spend their entire day in a self-contained classroom or even in a facility that is separate from the rest of the school system.

When working with the student with a mathematics-based learning disability there are a number of factors that must be examined to determine the most appropriate educational setting that is the least restrictive environment. These factors can be divided into two general categories: student characteristics and program characteristics. Student characteristics include academic level of functioning, behavioral level of performance, and learning styles and strategies. The program characteristics involve teacher receptivity toward students with special needs, organizational characteristics of the placement, and goals and objectives of the educational placement. Each of these factors will be examined as they relate to the student with a mathematics-based learning-disability problem.

STUDENT CHARACTERISTICS

Academic Level of Functioning

It is essential that school personnel be accurate in describing a student's level of achievement in mathematics. Typically, this description occurs using grade level equivalent scores taken from the mathematics subtest of an achievement test given to the student. (See Chapter 3 for a detailed discussion of assessment.) The student's overall performance in mathematics is based on the total number of items correctly answered. This number is translated into a grade level equivalent score according to norms developed by the test's author. The assumption underlying this approach is that students perform mathematics skills in a manner consistent with the sequence in which the skills are taught in the school curriculum. For example, fractional computations are usually taught after the student has mastered whole number operations. It is in third or fourth grade when many students are taught fractional computations. Therefore, a child found to be functioning at a 4.2 grade level equivalent in mathematics would be presumed to have mastered all the mathematical skills prior to that level. By inference, one would assume that the child had been taught and was competent in fractional computations.

This situation might be true for students of average learning ability. But, the LD student often shows large gaps in the skills acquired. Some students may know how to divide correctly a two-digit number into a six-digit number with a remainder and have difficulty with addition of four three-digit numbers. Other LD students may have a good understanding of geometry but be unable to reduce numbers to their greatest common factor. In short, grade level equivalent scores earned by LD students on achievement tests are often misleading because they do not describe accurately and in detail the specific mathematical skills mastered and those that are lacking.

There are at least two ways to establish the LD student's skill competencies in mathematics. The first is to examine the student's test record after taking an achievement test. Identify those items failed and give additional examples to see if the child merely blocked on the item at that point in time or is truly unable to solve the problem. If the achievement test given has a sufficiently representative number of mathematics problems this can be an effective and efficient way of pinpointing skill mastery and deficit. The key is the comprehensiveness of the test.

A more time-consuming way to identify areas of skill mastery and deficit is to assume that the LD student has not mastered any skills related to the topic being taught. The teacher then identifies those skills essential to success in that topic and constructs a criterion-referenced test to measure the student's level of skill development.

The greater the number of skills the student lacks the more seriously one must consider placement in a resource room or self-contained class made up of students with similar problems for mathematics instruction, especially at the middle school and secondary grade levels. At these levels the regular education mathematics program is often oriented toward conceptual development rather than teaching basic skills. Schools using tracking systems with lower and higher level classes are really grouping students together who have similar needs in mathematics instruction, but not necessarily a mathematics LD problem. At the primary and elementary grades the LD student is more likely to be successful if integrated within the regular mathematics program and given supportive or remedial help. This is because the content of the mathematics program is more homogeneous and defined than at the upper grade levels.

Behavioral Level of Performance

Mathematics is a subject area that differs from many other content areas in several substantive ways. It is a very disciplined area where answers are either correct or incorrect. The student can have no personal opinions or feelings about the sum of 22 and 16. Personal experience, verbal expressive ability, or literary talent will not compensate for lack of knowledge about mathematical operations and facts. It is only at the more advanced and abstract levels involving the philosophy of mathematics that these kinds of qualities become useful and relevant.

Because mathematics has a predictable and consistent body of knowledge the student learning mathematics must acquire a basic operational fund of information. Failure to acquire this knowledge prevents the student from performing in those areas where the knowledge is used.

Much of mathematics is also cumulative. When a child fails to master one set of information, the child may also be prevented from learning new concepts. This situation often occurs when teaching pre-algebra and algebra. If a student has not learned multiplication facts and understood what multiplication is, it will be impossible for the student to learn and understand algebraic operations.

Classroom structure and organization are prerequisites to mathematics instruction. Students who are not able to function within a classroom will be unable to learn mathematics. Students who are inattentive and distract others, lack self-control, or who simply refuse to apply themselves will fail. Alternative educational settings for mathematics instruction for these students must be given serious consideration. Not only will these kinds of behaviors interfere with the student's learning progress but they also can pose serious obstacles to learning for other students in the class.

Learning Styles and Strategies

Many LD students have problems in remembering and recalling. Some of these problems stem from the fact that many LD students do not learn as well as their average peers when taught using verbal explanations or listening to a lecture. Other LD children have problems learning when information is presented visually without an accompanying verbal explanation. It is not possible, however, to say that all LD children have the same kind or even similar kinds of learning styles and learning deficits. These characteristics vary according to the individual student.

Once it is determined which sensory modality is stronger for learning, a number of other learning style factors must be examined. Some LD children may understand fractional concepts only when they are taught using pictures and manipulatives. This challenges the creativity of the teacher to translate these concepts into their corresponding visual representations. Many LD children need more teaching presentations and practice trials than their peers to learn even basic mathematical operations. New concepts cannot and should not be presented until this student has mastered the preceding and prerequisite facts.

Another area where many LD students do poorly is in their ability to take information learned in school and apply it to other situations. This is most clearly illustrated at the secondary level vocational trade school. Within the mathematics class and the resource room the student may have mastered all necessary multiplication facts and be able to correctly multiply up to any four-digit by four-digit whole numbers. This student may also have mastered the concept of area and be able to calculate it when given a word problem. The student now comes to the shop class where the floor of a room is to be tiled. The teacher asks the student to calculate the square foot area for the room and to find how many 12 inch by 12 inch tiles will be needed to complete the job. The student is unable to do the calculations. Similar situations may exist in science, economics, or other classes where mathematics skills have an essential and basic function. Provisions in mathematics classes must be made to teach LD children how to use these skills in day-to-day real-life activities.

PROGRAM CHARACTERISTICS

Mainstreaming an LD child into a regular mathematics class is not simply a matter of complying with legal or school district mandates. Can the instructor sufficiently modify the curriculum to meet the diverse learning needs of one LD student? If four LD students are in the same class how much instructional and curricular modification is needed to meet their learning needs? Are their needs similar or totally different? What is the impact of changes in teaching or course content on the teacher, other students, and the overall quality of education within that particular classroom? Does the teacher know and understand specifically what the LD child's needs are and how to meet them? Are sufficient support systems and personnel available to assist the mathematics teacher with these students? These are only a few of the major questions related to learning styles and strategies that must be addressed before placing the LD child into a regular mathematics class.

Teacher Receptivity toward Special Needs Students

A major program characteristic to consider in determining whether to place an LD student into a regular mathematics class is the receptiveness of the teacher to work with this type of student. Some staff members may be reluctant to accept LD children in their classes because of expectancies based on either what they have heard from colleagues or previous personal negative experiences with these students. Some teachers expect that all LD children have some kind of brain damage that interferes with learning, while others might see them as being discipline problems in class. Some LD students do have either or both of these characteristics but many do not. Stereotyped attitudes on the part of teachers can have a great impact on the chances of success for any LD student placed in a given teacher's class or in other classes taught by colleagues of that teacher.

Some staff members may be reticent about accepting LD students into their classes because they feel that they lack sufficient knowledge and training in teaching methods and curricular adaptations to meet such students' learning needs. This is a realistic concern. School administrators must ensure that adequate inservice training opportunities, physical resources, and personnel are readily available within the school to prepare staff to deal with the mainstreamed LD student.

Teacher receptivity becomes especially important at the middle school and secondary grade levels where teachers often view their role as presenters of information to fulfill curriculum requirements established by the school administration. These teachers may be reluctant to reduce or modify the amount of content presented because of fear of criticism from their department head or principal for failure to cover the entire curriculum scope. If LD students are to be successfully mainstreamed into mathematics classes, teachers must be given explicit assurance by their administrators that failure to meet content requirements will not be viewed negatively. Meeting the needs of LD students in mathematics typically requires adaptation both within the curriculum and of the teaching pace. Although neither of these necessarily means that the total mathematics curriculum cannot be presented within expected time limits, the specific learning characteristics of LD students frequently restrict the extent of content to be taught.

Organizational Aspects

Although there are many possible classroom organizational characteristics that should be examined when placing an LD student into a regular mathematics class, three particularly important ones are the size of the class, the level of the curriculum, and the methods for teaching and learning generally followed in the class.

Many LD children become confused merely by being in a physically large room. Others become overwhelmed when asked to function with ten or more students in the same class. Each of these feelings and their accompanying behaviors occur because many LD students have difficulty mobilizing and organizing themselves. Sometimes these difficulties arise from feelings of inferiority and insecurity; at other times they arise from reasons unknown to either the child or professionals. In general, LD children need a high level of classroom structure and consistency. The larger the room or the greater the number of students in the class the less likely that the structure and consistency will be available to the degree necessary for optimum learning performance.

Clearly, mainstreaming in mathematics must take into account the academic levels of students and the curriculum. Placing an LD child into a mathematics class with three other students may meet that child's needs for structure and consistency but generate other problems if the three students are functioning either above grade level or have serious behavior problems. Even if the LD student is placed into a mathematics class at the correct grade level, the placement may be inappropriate if the class is comprised of students three years younger chronologically who are talented in mathematics.

The teaching pace for presenting information is also important. Some teachers tend to adhere rigidly to administrative timelines, while others proceed at the level at which most students in the class function. Neither approach may be responsive to the educational needs of the LD child.

Finally, the specific methods used by the teacher for instruction and the ways in which students are allowed to learn must be assessed relative to the LD child's needs. Some teachers prefer to lecture with students taking notes. Is the LD child an adequate auditory learner and are writing skills intact? Can the student distinguish between important facts and irrelevant details to take notes? Is the student sufficiently personally organized to maintain and retain a notebook during the school year? Other teachers may rely heavily on experiential learning activities. How do the various aspects of this approach correspond with the learning styles and educational needs of a particular LD child being considered for placement in this class? The closer the coordination between the LD student's learning characteristics and needs and the natural teaching and learning activities conducted within the mathematics class the greater the probability that the mainstream experience will be successful for both the student and the teacher.

Goals and Objectives of Educational Placement

The last program characteristic to consider is the goals and objectives for the mathematics course in which the LD child is to be placed. In general mathematics courses the goal may be to have students attain mastery of basic whole number operations. Is this a reasonable level of achievement for the LD student? Can deficiencies and gaps in whole number computations best be remediated on either a tutorial basis or in the resource room and the student placed into an introductory algebra class? If the LD student is to enter sometime after the school year has begun, what prerequisite mathematics skills must be made up? At some point during the LD student's educational career a group of professionals must decide whether the student is capable of college preparatory work or should be oriented more toward a vocational trade school or survival skills program. Each course has very different mathematics requirements. The desired learner outcomes for a particular mathematics course or class must be matched with the actual levels of achievement and expected levels of growth for the student.

PLACEMENT OPTIONS FOR SPECIAL EDUCATION PROGRAMMING

Once it is determined that an LD student can be educated at some level within the public school setting the next step is to identify the specific administrative arrangements between the regular and special education systems.

Table 11–1 indicates the variety of services that can be offered to LD students in the public schools. These services occur either within the regular school setting with or without special education or in another setting completely separate from the regular school setting. The model presented offers a sequenced organization of administrative and instructional placement options that relate to the identified needs of a special education student. This model also provides a graded structure designed to fit children into the least restrictive environment with the maximum benefit to both the child and school staff. Finally, it presents a general framework that integrates special education within the regular education system.

This placement model emphasizes the educational needs of children but also recognizes that some students who learn poorly in school do so because of

Table 11–1 Settings and Types of Services Available for LD Students

Level of Primary Class Setting	*Type of Service*
A—Regular Class Placement	• Special education consultation and inservice training as needed. • Special education consultation and inservice training on a regularly scheduled basis. • Specialized individual or small group tutorial services either during class or after school. • Resource room participation for one subject area only. • Resource room participation up to one-third of each school day for several subjects.
B—Special Class Placement Within Regular School Setting	• Resource room participation between one-third and one-half of school day. • Self-contained class on a half-time basis. • Self-contained class on a half-time basis with work-study. • Self-contained class on a full-time basis.
C—Alternative Class Placement Outside Regular School Setting	• Half-time special school placement with work-study. • Special school for day-long programming. • Homebound tutoring services. • Hospital or residential school.

noneducational reasons. Thus, the administrative responsibility for these programs gradually shifts from school to health or corrections agencies. As children are placed at higher levels within the model the placement reflects the increased severity and complexity of the problems and the inappropriateness of the regular class to address these problems in any reasonable manner. Higher levels of intervention or administrative placements separate children more and more from participation in typical daily school and home activities. The more specialized and segregated placement options require highly specialized staff and are more expensive to the school district.

In the area of mathematics instruction regular class placements either with teacher consultation or support from the resource room must take into account both student characteristics and curricular or program issues. While this model emphasizes the educational and noneducational needs of children, the nature of mathematics instruction is such that its contents and desired learner outcomes must be examined with a primary view toward the educational characteristics of the LD student.

THE LD STUDENT AT THE PRIMARY AND ELEMENTARY LEVELS

Criteria for Placement

A major emphasis in mathematics programs today is on encouraging the student to think independently and creatively in problem-solving tasks. Problem solving encompasses a variety of routine and sophisticated activities that are essential to daily living and to success in one's career or occupation. Children learn how to apply mathematics to real-life situations at many different levels of complexity.

At the primary and elementary grades problem-solving skills are taught as part of an overall approach where basic mathematical facts are presented. These facts include number identification, arithmetic computations, and classification as well as measurement, geometry, and inferential reasoning. An additional aim of the mathematics program at these grade levels is to stimulate interest and achievement, regardless of the ability level of the student.

It is in these early years that the teacher has the greatest degree of control over the child's success or failure in mathematics. At the primary level children enter on an approximately equal basis. Initial differences among students reflect family backgrounds or extreme physical, genetic, or emotional disorders. Gaps in the child's knowledge of mathematics are not due to personal aversion acquired for the subject because of failure or from inadequate teaching, differing teaching styles, or a number of other school-based factors that often become relevant at the upper grade levels. Variations in learning rate and progress among children reflect their individual characteristics in responding to the curriculum and specific teaching strategies used to present information. As we have noted the teacher must be sensitive to these individual characteristics and differences and be able to recognize when they are limiting achievement in the mathematics program.

One significant criterion in determining the need for additional help in mathematics is the child's achievement in relation to class peers. Is the child having problems acquiring specific concepts or is it a generalized learning disorder? The more generalized and extensive the learning problem the more likely that the child will need intensive special education intervention.

There are several approaches for use by the regular class teacher with the child having specific problems in mathematics. Teaching strategies might be modified to provide a greater number of concrete learning activities using the visual, auditory, tactile, and kinesthetic sense modalities in isolation and in combination with each other. Sometimes, just giving a student more opportunity for drill, practice, and repetition of concepts can increase the learning rate. Practice and drill should involve both paper and pencil tasks in addition to application of mathematical facts to real-life experiences. The greater the number of situations where the child can use mathematics the more likely that the concepts will be learned and retained.

It is important to recognize if a student fails to master a specific mathematics fact that this fact could be a prerequisite for success at some later point in the program. As such, a specific failure to learn at an early level could produce a more generalized learning deficit later on. Teaching must be oriented toward mastery learning with new concepts presented only after preceding concepts have been mastered.

The more severely learning-disabled student will show a pattern of overall and consistent difficulty in keeping pace with classmates. The gap between this student and his or her peers continues despite substantial program and instructional modifications made by the classroom teacher. The severely learning-disabled student may need so many teaching repetitions and opportunities for drill to acquire a concept that additional modifications within the regular class setting are unreasonable and impractical for the student, classmates, and the teacher. The child's specific learning styles and needs may be so unusual and complex that only the most highly individualized teaching program involving a high degree of structure and multimodal instruction is effective in producing learning gains. In either instance, the student is capable of learning what is being presented. The critical issues are the amount of time and the demands placed on the teacher for the child to attain a given level of performance. The student with the more generalized type of learning disability will require specialized assistance that probably can best be provided through placement in a part-time resource room or self-contained special class.

Selecting the Content of the Mathematics Program

The content of the mathematics program for the LD student at the primary and elementary grade levels should closely parallel that offered to students in the regular mathematics program. The major differences between the two programs will be in the pace at which information is presented and in the instructional strategies used.

Competence in basic arithmetic computations using both whole numbers and fractions is important. But of even greater importance is the child's understanding of the meaning of whole numbers and fractions and the processes involved in their operations. Rather than drill the LD student for hours on multiplication tables or subtraction flash cards it is recommended that the mathematics program emphasize comprehension of concepts and teach the student to become skilled in using a calculator to perform the actual operations. Many LD students can attain levels of understanding that are equal to those achieved by their peers. Given this situation and the ready availability of computing aids, it is inappropriate to focus mathematics instruction on the isolated memorization of routine computational facts at the expense of training in understanding and applying mathematics in problem solving. The development of functional competence in the student should be the mission of the mathematics program.

There must be a reasonable balance among instruction in basic computations, algorithms for use with simple problems, and the use of a calculator for more complex problems, particularly when computations can distract the student from the general principles presented in the problem. The young LD student must be given every opportunity to develop those skills and qualities that would be taught within the regular mathematics program, regardless of the classroom setting in which the instruction occurs. At a minimum these skills should include problem solving, applying mathematics to daily life situations, estimating and approximating, basic arithmetic computing, measuring, interpreting graphs and tables, and using a calculator. In addition, all LD students should have the opportunity to learn to use the microcomputer or minicomputer as part of their mathematics program.

THE LD STUDENT AT THE SECONDARY LEVEL

Criteria for Placement

The overall school situation for LD students at the secondary level is quite different from that at the primary and elementary grades. Teachers often see themselves as content area specialists whose role is to disseminate information about a given subject. There may be a general reluctance to modify curriculum or instruction to meet the needs of a few selected students. Modification may be seen as dilution of standards within the field. These teachers may view educational remedial efforts as the responsibility of someone else on the staff.

Each added year of schooling brings about increased levels of expectation by teachers for students. At the secondary level teachers make assumptions about what prerequisite knowledge a student should have in a subject area. The curriculum is then developed according to these expectations. The student who has not mastered the necessary skills, for whatever reasons, is at a clear disadvantage.

Another major difference between the secondary and earlier grades relates to the attitudes of students about school in general and mathematics in particular. Younger students typically enter school with enthusiasm and eagerness. As the LD student experiences failure and personal frustration because of an inability to keep pace with peers, school becomes increasingly associated with negative and unpleasant experiences. This situation is magnified in the case of mathematics instruction. Mathematics class is often an anxiety-inducing experience even for students with average ability or higher. For the LD student mathematics class can be especially threatening. Consequently, the secondary level teacher must now address the student's apprehensions and feelings about mathematics and school in addition to teaching the content of the curriculum.

The primary criterion to determine if a student needs tutorial or remedial help external to the mathematics class is to compare that student's learning rate and level of achievement with class peers. Is the overall achievement level about consistent with that of the class? If it is consistent, are there specific areas or concepts with which the student is having difficulty? Once again, the more generalized the learning problem the greater the need for intensive and specialized remedial assistance.

At the primary/elementary levels the major thrust of intervention efforts often can occur in the regular class setting coupled with either consultation from the special education teacher or that teacher working directly with the student. Similar organizational arrangements can be used in the secondary level mathematics program to more fully meet student needs. LD students from several different classes with similar problems in learning either specific concepts or in their total understanding of mathematics might be grouped together (not to exceed ten students) and given instruction by a teacher from the mathematics department. The major characteristic of the class would be the teacher's flexibility in varying and adapting the mathematics curriculum and the teaching strategies and materials to present the information. This teacher could be given additional planning time during the week to allow for the extra preparation time essential to offering the most appropriate program of mathematics instruction to these students. LD students with more intensive or complex educational needs could receive their mathematics program in a special setting such as the resource room, a part-time self-contained class, or a mathematics laboratory. Each of these settings would contain specially trained personnel who could more fully meet student needs through smaller group (not to exceed five students) or individual instruction.

Within either organizational arrangement the amount of time actually spent by students could vary according to individual needs. Two options exist. First, the severely learning-disabled student in the self-contained class could spend part of the mathematics class in the larger part-time class. Second, selected students in the larger class might spend part of their mathematics class in a regular mathematics program depending upon the specific concepts being taught. Several other time arrangements are also possible, depending on student needs, characteristics, and the mathematics curriculum offered.

The specific numbers of students in a given mathematics class will vary according to the amount of attention that must be given to achieve desired learner outcomes. Instruction can be individualized using either computer-assisted teaching materials or practical experiments and projects that allow for independent study or small group activities. One final possibility is to use students from upper level mathematics courses to tutor LD students. This must be done carefully so that the dignity of the LD student is maintained within the peer group.

The individual teacher of mathematics must be creative in varying teaching methods, class groupings, materials, and the curriculum to best fit student characteristics and needs. Using alternative activities that involve similar mathematics content and operations is a way of achieving variety within the context of practice and drill in an applied manner.

Selecting the Content of the Mathematics Program

The typical high school mathematics program presents a variety of course offerings and levels of complexity within many courses. Course offerings range from remedial and general mathematics through business or applied shop mathematics to analytic geometry and calculus. It is reasonable to assume that many LD students are capable of achieving successfully in many of these courses, even at the most advanced levels. For LD students to attain these levels of achievement it is often necessary that teaching approaches be adapted to present information in a relevant, meaningful, and comprehensible manner.

In deciding what specific content and level in mathematics to present to the LD student two concerns must be addressed. The first has to do with essential skills. The second involves overall career preparation or occupational skills.

Essential Skills

Technological advances across many areas have created a growing diversification in the applications of mathematics to everyday life. Even a task as relatively uncomplicated as shopping for groceries involves a number of mathematical operations. First, what proportion of the weekly salary will be spent for food? Is this amount adequate to purchase those items essential for a reasonably balanced diet? Does this proportion leave enough money for other bills and needs? Will the food bought with this money last the entire week? How does one get the best value for the money spent? Is it better to buy a 12-ounce jar of mustard for 79¢ or a 16-ounce jar for 93¢? Is it wiser to use a 25¢ coupon to buy one loaf of bread that weighs 32 ounces and costs $1.09 or to buy two 24-ounce loaves on special sale for $1.25? How do the purchasing advantages relate to the rate of consumption or the time period during which the items will be stored? All students should be taught the necessary skills so they are capable of successfully making these kinds of decisions using mathematical reasoning. With the LD student it is particularly important that a mastery learning approach be used to ensure that no new skills are taught until skills being taught are mastered and can be applied consistently.

This kind of programming requires recognition of the diversity of needs among students. Small classes and alternative student groupings using some of the guidelines discussed earlier are organizational ways of dealing with these needs in a reasonable way that does not overwhelm the teacher. Other content areas in daily living where mathematics is important include reading timetables and schedules, interpreting graphs and tables, using a hand calculator, reading measuring instruments, maintaining savings and checking accounts, reading newspaper advertisements and the sports pages, and developing appropriate health and dietary habits. Some other areas that will be useful for students on a longer-term basis are interpreting the stock market, purchasing a car or house, calculating energy costs, buying insurance, and computing taxes.

Not all LD students will attain competence in each of these content areas. But significant numbers will be successful provided they are given sufficient opportunity for learning in a highly flexible curriculum that stresses the practical applications and uses of mathematical operations and reasoning.

Occupational Skills

Although much of the following discussion will address mathematics as related to vocational preparation, it is essential to note that career preparation is not necessarily limited to formal vocational-technical education that concludes at graduation. Career preparation can extend up to and beyond college into graduate and professional school.

Vocationally oriented mathematics courses show students the practical and important uses of mathematics as a tool to perform a job well. These courses must take at least three different perspectives. First, the teacher must review each student's skills in basic operations. Students lacking these skills must be brought up to expected levels of performance. Second, the teaching program must focus on practical uses of mathematical concepts and skills as related to the job. Finally, as students develop increased skill they may need to be exposed to more advanced mathematics related to the job. In these courses there is little tolerance for student failure or even for attaining mastery less than 100 percent. The carpentry student who is able to measure to ¼ inch with only 80 percent accuracy will make a very poor finish carpenter whose work affects both the employer and customer in terms of money, time, and quality.

In developing this type of course much of the specific content and principles to be taught will come from the vocational education teacher. This teacher's first task is to define those mathematics skills that are essential to job success and those with only secondary importance. Once the specific skills are defined, the next step is to integrate these skills within the mathematics program. This requires close coordination between the mathematics and vocational teachers. The mathematics teacher can give the vocational teacher information about those skills the student has mastered and lacks. The vocational teacher can offer information about assignments, tasks, and experiments where the student can use the skills being taught. Worksheets and homework assignments can be mutually developed and used in both classes. The sequence for this coordinated approach to relate mathematics to other subject areas is as follows:

1. Vocational teacher lists mathematics skills for each unit to be taught and gives the lists to the mathematics teacher.
2. The mathematics teacher develops a program of instruction focusing on these skills and the prerequisite ones.
3. Both teachers jointly develop classroom and homework assignments, tasks, and experiments to give the student practice and variety in their application.
4. Each teacher evaluates the learning rate and progress of the student and reports this data to the other.
5. Both teachers adjust instruction and curriculum to meet the learning characteristics and needs of students.

This approach provides diversity in the kinds of tasks given to the student and allows for drill in using mathematical operations and concepts in real-life settings. Exhibit 11–1 presents an example of an assignment developed using the above

Exhibit 11–1 Developing Coordinated Assignments in Mathematics

Needed Mathematics Skills
1. addition of mixed numbers
2. knowledge of width and length

Problems: To find the ROUGH OPENING (RO)
To find the HEADER LENGTH (HL)

Solution Steps
1. Find RO according to *door type*
   - Hinged, single swing, and accordian type doors
     1. Find door size (width is first, height is next)
     2. Find floor type
        Wood floor—add 2″ to width
        add 3″ to height
        This is the RO: ________
        Concrete Floor—add 2″ to width
        add 2″ to height
        This is the RO: ________
   - Bipass and slider type doors
     1. Find door size
     2. Find floor type
        Wood floor—add 1″ to width
        add 3″ to height
        This is the RO: ________
        Concrete floor—add 1″ to width
        add 2″ to height
        This is the RO: ________

Exhibit 11–1 continued

2. Record RO
3. Find HL by adding 3″ to width for the RO.

RO and HL Worksheet Assignment

NAME ______________________ DATE ______________________

Find the RO and HL for these doors:

Door Type	Floor Type	Door Size	RO	HL
Hinge	Wood	2′0″ × 6′8″		
Slider	Concrete	2′4″ × 6′8″		
Bipass	Concrete	4′0″ × 6′6″		
Swing	Wood	1′10″ × 6′7″		
Slider	Concrete	4′10″ × 7′2″		
Accordian	Wood	2′2″ × 6′6″		
Single Swing	Wood	2′4″ × 6′8″		
Hinge	Wood	2′0″ × 7′0″		
Hinge	Wood	2′2″ × 6′6″		
Bipass	Wood	4′1½″ × 6′4¼″		
Single Swing	Concrete	2′4″ × 7′0″		
Slider	Concrete	5′8¼″ × 7′1¼″		
Hinge	Concrete	2′0″ × 8′0″		
Accordian	Wood	5′4″ × 6′10¼″		
Accordian	Concrete	3′2″ × 5′9½″		

approach for an intermediate level carpentry class with moderately learning-disabled students. The assignment was followed by the students actually calculating the RO and HL on in-class models and then on a real house that the class constructed.

Finally, this approach allows for careful monitoring of the effectiveness of instruction and the student's learning rate. It shows students the practical uses of mathematics and the reason for learning how to use mathematics. Through necessity, many vocational mathematics courses will have to address the content found in geometry, trigonometry, and algebra. This is especially true for vocational-technical trades such as carpentry, automotives, and building and grounds maintenance.

The crux of the program of mathematical studies is to teach students problem-solving strategies and clear reasoning. This goal applies whether the program is

Exhibit 11–2 Situational Worksheet To Teach Problem-Solving Processes

The Problem:

You are to tile the floor of a room that is 12′ × 14′. You will be using floor tiles that are 12″ × 12.″ What are the steps you would use to find how many floor tiles to buy to tile the floor of this room? ______________________________

____

If the tiles cost you 73 cents each, how would you find the cost to you as the contractor for just the floor tiles?

____

____

Are there other expenses or costs you should add in to the price you would charge to a customer for tiling this room? What are these added costs?

____

____

What are these materials used for in floor tiling?

____

____

NOW, find the square foot area for this room. ______________
How many tiles will you need for the room? ______________
What is the cost for the tiles? ______________
What is the total cost to you as the contractor? ______________

oriented toward life skills or career preparation. With the LD student this will often mean focusing on both the processes used to solve a problem and the actual solution or answer. This dual focus can be addressed by using situational worksheets such as that presented in Exhibit 11–2 and asking the student to identify only the steps needed to solve the problem. A specific format must be arranged based on the severity of the student's learning disorder. More severely learning-disabled students may be asked to identify only the first step or two before being given the total problem. This controlled rate of presentation teaches the student how to sequence information correctly and to organize the cognitive problem-solving approach in a logical way. Less severely learning-disabled students may be able to handle the problem-solving process as a total entity and not need the process segmented into smaller units of information. For both groups the emphasis is on what you do to get a solution.

Several instructional options can be used with this approach. The teacher can present an array of answers in a multiple-choice format and ask students to order them correctly. The task could be left more open ended with the teacher using a graded point system so that no student can fail. More points are given for answers that are based on precision and the correct application of mathematical processes and operations to arrive at a solution. (This could be easily used in the task presented in Exhibit 11–2.) Word problems should be phrased in a way that leads to the discovery and application of principles.

The instructional tactics we have discussed can be used in nearly all subjects and most secondary level mathematics courses, including algebra, geometry, and trigonometry. Once the teacher has developed the necessary content for the mathematics program, the materials can be used with minimal updating or redesigning. The teacher now becomes free to attend to learning styles and strategies.

Index

A

Note: Page numbers in *italic* indicate entry is found in an Exhibit, Figure, or Table.

B

C

D

E

F

G

H

I

N

Q

R

Y

Z

About the Editor

John F. Cawley is Chairperson, Department of Special Education, University of New Orleans. He is especially concerned with curriculum and instruction for the handicapped.

About the Contributors

Colleen Blankenship is Associate Professor, Department of Special Education, University of Illinois. Her primary interests are data-based instruction and mathematics for the handicapped.

Anne Marie Fitzmaurice-Hayes is a mathematics educator with extensive research in curriculum development for the handicapped. She is presently Assistant Professor at the University of Hartford, where she teaches courses in mathematics and special education.

Henry Goodstein is Associate Professor of Special Education, Department of Special Education, Vanderbilt University. His special interests are assessment and curriculum and instruction in mathematics for the handicapped.

Elizabeth McEntire is Assistant Professor of Special Education, University of Alabama at Birmingham. Her areas of specialty are learning disabilities, classroom management, and behavior disorders, with special attention to mathematics.

Mahesh C. Sharma is Director of the Center for Teaching/Learning of Mathematics, Framingham, Massachusetts. Dr. Sharma also directs the Mathematics Institute at Cambridge College, Cambridge, Massachusetts. He has written extensively in the area of learning disabilities and mathematics and is noted for his client-centered efforts.

Robert A. Shaw is Professor of Mathematics Education, University of Connecticut. He has worked extensively in the area of mathematics and the handicapped and has been active in the curriculum development process.

Raymond E. Webster is Assistant Professor, Department of Psychology, at East Carolina University. Prior to this position, he served as a teacher, psychologist and administrator for 13 years. He has published extensively in the areas of information processing, mathematics, and the handicapped.

About the Contributors

X0046798 5